DIALOGUES

Stanisław Lem Titles from the MIT Press

Dialogues, translated by Peter Butko
Highcastle: A Remembrance, translated by Michael Kandel
His Master's Voice, translated by Michael Kandel
Hospital of the Transfiguration, translated by William Brand
The Invincible, translated by Bill Johnston
Memoirs of a Space Traveler: Further Reminiscences of Ijon Tichy, translated by Joel Stern, Maria Swiecicka-Ziemianek, and Antonia Lloyd-Jones
Return from the Stars, translated by Barbara Marszal and Frank Simpson
The Truth and Other Stories, translated by Antonia Lloyd-Jones

DIALOGUES

STANISŁAW LEM

translated by Peter Butko

THE MIT PRESS
CAMBRIDGE, MASSACHUSETTS
LONDON, ENGLAND

© 2006 Tomasz Lem
English translation © 2021 Peter Butko

This publication has been supported by the ©POLAND Translation Program.

All rights reserved. No part of this book may be reproduced in any form by any electronic or mechanical means (including photocopying, recording, or information storage and retrieval) without permission in writing from the publisher.

This book was set in Dante Pro and PF Din by Jen Jackowitz. Design by Marge Encomienda. Printed and bound in the United States of America.

Library of Congress Cataloging-in-Publication Data

Names: Lem, Stanisław, author. | Butko, Peter, translator.
Title: Dialogues / Stanisław Lem ; translated by Peter Butko.
Other titles: Dialogi. English
Description: Cambridge, Massachusetts : The MIT Press, [2021]
Identifiers: LCCN 2020047225 | ISBN 9780262542937 (paperback)
Subjects: LCGFT: Essays.
Classification: LCC PG7158.L39 D513 2021 | DDC 891.8/5473—dc23
LC record available at https://lccn.loc.gov/2020047225

10 9 8 7 6 5 4 3 2

CONTENTS

Preface to the Second Edition   vii

Dialogues

I   3

II   21

III   31

IV   47

V   63

VI   99

VII   115

VIII   163

Supplement 1: The *Dialogues* Sixteen Years Later

Lost Illusions, or From Intellectronics to Informatics   193

Applied Cybernetics: An Example from Sociology   219

Supplement 2: Additional Essays

The Ethics of Technology and the Technology of Ethics   241

Biology and Values   283

Translator's Notes   335

PREFACE TO THE SECOND EDITION

This book, written between 1954 and 1956, published in 1957, and conceived under the spell of cybernetics, may be occasionally dated on the level of facts; on the level of making predictions, the passage of time has either corrected it or shown it to be false.

These words hardly look like a recommendation for a text that was designed to popularize the fundamental concepts of and foresee the future development of cybernetics. Bringing the book up to date has proved impossible—adhering to its original purpose, I would need to write it anew. Yet I have not changed a word of it, adding a supplement instead, because when I reread the book, I realized that it preserved a quality different from the one I originally intended. Time has made *Dialogues* a testimony to the almost limitless cognitive optimism that the emergence of cybernetics created in me. The book describes not so much cybernetics or its foundations as the visions it offered at the end of the 1950s—visions that were not only my own. The supplement begins with two critical sketches: the first briefly summarizes the historical journey of cybernetics; the second is a contribution to the cybernetic theory of the sociopathology of governing. In the first, the views expressed in *Dialogues* are confronted with the actual state of things after the sixteen years that passed since the book was written. The naivete revealed by that confrontation was not only mine; it was widespread among the proponents of cybernetics in the 1950s. The comparison of past opinions with present realities makes an interesting contribution to the history of science. It nicely illustrates the extrapolative linearity that marks almost all scientific revolutions. To those at the moment of a revolution, the prospects in the development of knowledge appear straight, as if the complex, winding path, full of obstacles and blind alleys, that led to the revolution was suddenly replaced by a smooth exponential trajectory without obstacles or reversals. The invariable result of this divergence between overoptimistic hopes and reality is a subsequent cognitive pessimism, but such pessimism is usually based on a misunderstanding. Cybernetics may not have accomplished what was most eagerly hoped from it—in particular, it did not cure the plague of specialization (this was expected because cybernetics is an inter- or maybe even superdisciplinary science, unifying natural sciences with the humanities)—but it accomplished what no one thought of. Although admittedly computers have not become equal companions to people, they have proved irreplaceable tools in regulating the world economy. Although information theory has not become the new philosopher's stone, it has spread into fields where it was totally unexpected—for example, theoretical physics. We could list many other instances of such divergence between expectations and actual accomplishments.

So I felt that my supplement would have some cognitive value, particularly today, when we so often see fashionable futurological concepts blossom, concepts that tend to reduce themselves to a dense proliferation of prognostications, which are disproved or ridiculed in just a few years after their birth, and yet this phenomenon is conveniently ignored by the adepts of futurology—to their field's detriment. (A quick look, for example, at what Hermann Kahn had to say about the world's global political matters in his book, *The Year 2000*, written with A. J. Wiener in 1967, is enough to prove that everything that has happened since is *totally* different from all those "canonical," "noncanonical," or "alternative" predictions of his lengthy volume.) Knowledge is unquestionably a more important matter than the reputation of a futurologist. Therefore, comparing the prognostications and opinions of the relatively recent past with those of today may be instructive for us.

The second part of my supplement contains a sketch that amends the last paragraphs of *Dialogues*—about the pathology of social regulation. My comments here are those of a nonexpert; what motivated me to include them in the book was that the second part of *Dialogues* had not dated as much as the first.

Finally, I added two short discourses that were published in *Studia filozoficzne* (*Philosophical Studies*)[1] and are thematically linked to the book's overall thesis. The first talks about the ethics of technology and the technology of ethics; the second deals with the problem of value in biology. The first considers the questions of order in society, civilization, and ethics; the second, the relation between axiology[2] and theoretical biology. I include these essays in this edition because *Dialogues* was conceived not as a tractate, separated into two voices, on the new scientific discipline and its possible future development, but rather as a search for the investigative methods and tools that might increase our understanding of both the human and nonhuman worlds. Cybernetics was viewed from the perspective of its possible applications, not as a "pure" science like mathematics (as some scientists wished to see it). That *Dialogues* eventually turned out to be an expression of the cognitive curiosity and anxiety of modern thought is my justification for putting both these studies in the Annex.

Kraków, December 1971

DIALOGUES

I

PHILONOUS[1]   Hello, dear friend. What are you pondering, all alone, in this beautiful park?

HYLAS   Oh, it is you! I am glad to see you. Last night, an idea occurred to me that would greatly benefit humanity.

PHILONOUS   What is this wonderful idea?

HYLAS   I became convinced that people will achieve immortality in the future.

PHILONOUS   Do my ears deceive me? Is it possible that you have turned your back on the materialism that you always professed?

HYLAS   Not at all. My idea does not clash with materialism in the least; on the contrary, it inevitably follows from it.

PHILONOUS   Most interesting. Continue, my friend.

HYLAS   As you know, nothing exists except matter. These clouds, trees, the sun that is turning yellow, and the two of us are all material objects, that is, collections of atoms. The different properties of bodies are the result of differences in their atomic structures. Because the atoms of oxygen, carbon, and iron in the rocks, in the leaves, and in our blood are all the same, the rocks, leaves, and blood differ only in the distribution of these atoms, that is, in structure. It is the most universal truth that there are only atoms and structures made out of them. So I was thinking, why is it that despite the passage of time I see myself as the same Hylas who used to play here as a boy? Can this sense of personal identity, I asked myself, be due to the persistence of the building blocks of my body, that is, of the atoms that compose it? But that is impossible. We know from the natural sciences that the atoms in our body are continually being renewed from the food and drink that we ingest and from the air that we breathe. What, then, is the source of this sense of identity? Without a doubt it can only be the persistence of the structure that the atoms make. Because the atoms currently in my body are not the atoms that were there a month ago; they are only the same kind but form the same structure—and this is entirely sufficient. I conclude, therefore, that it is the persistence of my structure that guarantees the continuity of my existence.

PHILONOUS   So far, so good. Go on.

HYLAS   In the future, people will be able to copy atomic structures of physical systems with increasing fidelity. Even now they can make artificial diamonds and sapphires, and synthetic urea and proteins are synthesized in a test tube. Eventually people will master the art of building the molecules of a living body and the body itself—out of atoms. Then they will attain immortality as they will be able to bring anyone back to life by exactly assembling atoms to match the same structure that a body possessed before its death. I envision this resurrection process will take place in a machine provided with the blueprint or map or structural pattern of the deceased person. The machine will build the protein molecules, cells, tendons, nerves—and the person will step out, alive, smiling, in the pink of health. What do you say to that?

PHILONOUS   I say that we need to look at this problem from more than one angle.

HYLAS   What other angles are needed? We cannot construct such a machine yet, but scientific progress assures us that it will be constructed one day, and to us philosophers it does not matter if that happens a thousand or a million years from now. As I have said, there is nothing in nature but atoms and the structures made from them. Especially there is no eternal soul that would fly from a dying body to an afterlife. Whoever masters the art of assembling atoms to precisely match the structure of a body long in the grave will be able to call that person back into existence. Composed anew out of the individual atoms, the body even of someone buried centuries ago will stand before us, full of life.

PHILONOUS   Is this what you think? Good. To determine the specifics of your machine that resurrects with atoms, will you permit me to pose a few questions?

HYLAS   Certainly.

PHILONOUS   Excellent. Now imagine, Hylas, that you are going to die today, for you are in the hands of a tyrant who is determined to take your life and has the power to do it. The time of your execution has been set at seven in the morning. At six, which is actually now, you set out for your final *ante mortem* walk, filled with sorrow and dread. You and I meet, and you tell me of your misfortune.

Let this serve as an introduction to our hypothetical discussion: you are condemned to death, and I, your friend who desires to help you, am also the inventor of the atomic resurrection machine.

HYLAS   Fine. Continue.

PHILONOUS   My poor Hylas, you are going to die, how awful! But you are a materialist, yes?

HYLAS   I am.

PHILONOUS   Perfect. I have constructed the machine about which we have just been speaking. The copy of a person that I can produce with it is indistinguishable from the original, down to the last atom. It possesses not only the same body but exactly the same mental features, such as memory. As you know, memory resides in specific structural features of the brain. My machine's copy preserves all the details of the brain's structure and therefore all the memories of previous events, all the person's thoughts and desires. So when you die an hour from now, I will turn on the machine before your corpse turns cold, and from the same atoms that make your body now it will build a living and thinking Hylas. Without fail. Are you content?

HYLAS   Totally, of course. All you need to do is determine my atomic structure so that you can feed it into the machine.

PHILONOUS   Quite. But I would like to reassure you even more, dear friend, that you will survive your death. You know me and trust my word, but human-made things sometimes fail. Allow me, then, to create the Hylas who is to be your continuation right now. After your death, he—that is, you—and I will together enjoy the pleasures of life regained.

HYLAS   Philonous, what are you saying?

PHILONOUS   To make sure that nothing goes wrong, I will create the copy of you now . . .

HYLAS   But that's nonsense!

PHILONOUS   How so?

HYLAS   The copy will be not me but an altogether different person!

PHILONOUS   You think so?

HYLAS   He may be so like me that everyone will take him for me; he may have the same feelings, tastes, and hobbies, and be able to finish the work I have started exactly as I would, yet he will not be me. He will be a double, or twin, while I will be no more.

PHILONOUS   Why are you so sure?

HYLAS   Because if you create him now, and he is present, I will refer to him as "he." He will be outside me, distinct from me, another human being, and the fact that the two of us are like two peas in a pod will not make my death the least bit easier. For those who keep on living, for my friends and family, the illusion of my continuing existence will be perfect, but I—I will be dead forever.

PHILONOUS   How can you be so sure?

HYLAS   Can there be any doubt? Surely you are just provoking me, Philonous. If I picked up this wet and wilted leaf from the ground and handed it to "the other Hylas" standing here, it would be he who smelled its bitter but pleasant odor, not I. And it would be the same after my death, because my death would not affect or change him; he would continue walking through the world and enjoying its beauty after I ceased to exist.

PHILONOUS   Really? A problem. What should I then do with the machine so that your resurrection is guaranteed?

HYLAS   It's simple. Create a living and thinking copy of me *after* my death.

PHILONOUS   Is that what you think?

HYLAS   Yes.

PHILONOUS   A copy created *after* your death will be you, while a copy created before your death will not? What exactly is the difference between these two Hylases?

HYLAS First, the one created before my death will see me, and I will see him; second, he will know that I died and that he was created; third . . .

PHILONOUS If that is the difficulty, I can easily solve it: the copy will be kept in a deep sleep through medication; waking only *after* your death, it will not know the misfortune of your execution or of its own coming into the world.

HYLAS That was not my point. I can see, Philonous, that I must word things more carefully. Let us approach this problem with the sharp tool of reason. At the moment of my death, when I have ceased to exist, my copy will have no way of knowing that he is not me but merely a copy. True?

PHILONOUS True.

HYLAS But if you create the copy earlier, it is obvious that he is not me, because he exists independently of me, occupying a different space. His parallel existence *ipso facto* rules out a continuation of me. Now I see my error. The copy created after my death is me; the one created before my death is another, separate person. From which it follows that just one-millionth of a second would make the difference between my resurrection and the creation of someone who is a stranger to me, though my double, however bizarre that sounds.

PHILONOUS So you say that a copy of you assembled from atoms before your death will be a person who has no connection to you apart from an extraordinary similarity. Whereas a copy of you created after your death will be your continuation, that is, yourself. Right?

HYLAS Yes.

PHILONOUS And the difference between the two copies is what?

HYLAS The time of their creation. The parallel existence of the copy and myself rules out a continuation, but the copy's existence after the moment of my death makes my continuation possible.

PHILONOUS OK. Now I will tell you the kind of death the tyrant has decided for you. You shall drink a bowl of deadly poison. The agony will last an hour. When should I switch on the machine?

HYLAS When I completely cease to exist.

PHILONOUS When I make the copy at that moment, it will be your continuation, correct?

HYLAS Yes, if it is created after my death.

PHILONOUS Good. What if, just when you have lost consciousness due to the poison's effect, the devious tyrant orders his physicians to revive you by pouring an antidote into your esophagus through the rachis of a goose feather? The copy created by the machine after your death, as you have said, will be you. But now, because of your revival in the execution chamber, will that copy suddenly stop being you and turn into a totally separate person?

HYLAS How can I be revived after I have died?

PHILONOUS That is undoubtedly easier than constructing an atomic resurrection machine. Are we discussing technical or philosophical issues here, dear Hylas? Is there any principial, fundamental obstacle to bringing back to life a freshly deceased person? Are not surgeons capable of reviving the clinically dead on the operating table already today? Explain to me, please, what happens to the copy that has already been declared your continuation the moment you are restored by the antidote. Is it possible that it is not you but someone else who wakes in your original body?

HYLAS No. Of course it is I who wakes in the body that the tyrant killed by means of poison. Consequently, the copy ceases to be a continuation of me.

PHILONOUS Really? Consider, Hylas. Imagine that you are the copy, and I am showing you the machine and telling you that you just stepped out of it. Of course you perceive yourself to be Hylas in every inch of your body—as the machine did an excellent job. Imagine further that "the other Hylas," in the tyrant's captivity, was poisoned an hour ago but the physicians restored him with an antidote only now. Will you undergo any change in personality as a result?

HYLAS I will not.

PHILONOUS See, the copy is an ordinary living person (the machine works) who cannot be changed by what happens to the "original." Whether the "original" is served a poison or an antidote does not affect or change the copy at all. We can conclude that since no causal link exists between you, in your plight, and the Hylas created by the machine, say, a year from now, whether you are still alive or dead, he will be a distinct person, having nothing in common with you apart from the astonishing similarity. Your parallel existence with the copy, we agree, makes the continuation impossible. But whether the copy created after your death is truly you, thus providing the possibility of resurrection, is yet to be determined. So far, everything points against that conclusion.

HYLAS   Hold on. You confuse the issue. This is my body. When it is destroyed and has died, a similar structure may emerge in the future and . . . Ah, now I get it! We need to delay the copying until my body has entirely ceased to exist and its structure is totally gone.

PHILONOUS   So now it is the extent to which your corpse has decayed that decides whether or not you are resurrected? Resurrection then depends on the rate of decomposition of your remains. So if the tyrant has you embalmed, you can never come back to life. Is that what you mean?

HYLAS   Heavens, no! I see now that we need to separate our argument from the living beings. When people are involved, an element of fear or anxiety always interferes. Let us limit ourselves to inanimate objects. I have here a precious ivory cameo. Assume that I break it into atoms and then create an exact copy of it from those same atoms. What conclusion can we make? I see it like this: If we agree that the copy is a continuation of the original, then it will be just that. And if we say that it is not a continuation, then it will not be. It is just our agreement that decides, since no examination of the "earlier" cameo and the "later" one can possibly reveal any difference—because *ex definitione* they are identical.

PHILONOUS   Finally, you are shedding light on the problem. Applying your conclusion to a person would go like this: When you die at seven and I re-create you from atoms, your continuation depends on what we agreed beforehand. But isn't this absurd? If you die under a surgeon's knife but the doctors revive you with their skills, will you still say that your continuation depends on what we agreed beforehand?

HYLAS   The difficulty, as I see it, is this. For all objects that exist around me, continuation after atomic reassembly depends on an arbitrary agreement or convention. But when it is I who am reassembled in that way, this convention leads to an absurdity. Why this is so I cannot fathom. Surely a human being is no less material a thing than a piece of rock, a tuft of yarn, or a chunk of metal!

PHILONOUS   I will reveal to you the source of this dichotomy. When we set out to prove a continuing existence of an object, we also choose concrete criteria to determine if this is the case, that is, if the object before is the same as the object after. Thus we implicitly (but sometimes also explicitly) select the methods that we use to make the decision. However, I experience my consciousness directly and cannot choose a method to examine whether or not I am conscious at any given moment. Other people may view me as an object and easily establish criteria to decide the question of my continued existence after death and atomic re-creation, but I cannot. This is a universal problem of methodology. An object may manifest various properties depending on the method we use to examine it. But human consciousness manifests itself to its possessor in a way that is the most direct, primary, and self-evident, and requires no method or, if you prefer, uses "the one and only" universal method that all conscious and sane people utilize. We may have serious questions about the structure of consciousness and the mechanisms by which it arises, but no one questions its uninterrupted existence within any given individual.

HYLAS   Well, I think you are posing fallacious questions. The whole problem is set up incorrectly, as an *argumentum ad hominem*. You ask me about future events that I only can imagine because no one has ever experienced them. Furthermore, what matters here is only what I say *ante mortem*, because when you ask the machine-made copy after my death, he will of course say that he is Hylas, the very same who has had this conversation here with you. Everything that I have said to you on the topic of my future after my death and reassembly from atoms, and especially about the continuation of my consciousness, is subjective, what I imagine, expect, think, feel, and doubt, nothing more.

PHILONOUS   What? So the machine does not bring the dead back to a renewed life?

HYLAS   I am not saying that. I don't know the truth. But scientifically, nothing can be proved here. No conclusive experiment can be conducted, because the identical copy, when questioned, will always claim to be me, and there is no way—there cannot be—to tell that it is not merely a double. Thus standing on the firm ground of empirical sciences, one must conclude that the whole question is bogus, now and forever: any statement from me or anyone else testifies only to certain peculiarities of the human mind and can shed no light on what happens in the future. When it seems that it is doing so, it turns out to be just a misuse of language and nothing more. I am sure now, Philonous, that there is no real problem here.

PHILONOUS   You are right that the problem cannot be solved empirically. Even if the machine stood in front of us right now and you agreed to submit to the experiment and were annihilated and returned to life inside the machine, we would not know if it was you who were resurrected or if it was just a facsimile, a twin of you. What we face here is a rigorous either-or statement in logic: either the copy is a continuation of the original or it is not. If one of the alternatives leads to a logical contradiction, we must reject it and we declare that the other corresponds with reality. At any rate, I do not believe that our problem is fallacious. A fallacious problem is one that in fact does not exist. And if our problem does not exist, you have no reason to worry about what is going to happen at the seventh hour on the cruel tyrant's orders.

HYLAS   Philonous, you are joking about a problem that is difficult and should be analyzed matter-of-factly. Having been condemned to death, I do worry, because the announced execution is a fact that will happen and not a fallacious problem, while the possibility of resurrection, which I previously considered a certainty, remains shrouded in mystery. Let us apply the problem to a person. Suppose there exists a man Ex and that, while he is alive, we make a copy of him, Ex-Prime, in the machine. Ex and Ex-Prime both have the same perception of their identity and the same memory. When questioned, both will say that they have experienced the same past, although in fact only Ex has experienced it. Human identity therefore depends not only on the body's atomic structure but also on the genesis link between it and the structure that preceded it. The inclusion of this genesis element has saved our concept of identity for a future examination. Following Lewin, we can name this concept *genidentity*.[2]

PHILONOUS   I listen with pleasure to your reasoning, my friend, but I think it contributes nothing to the topic of our discussion, or, worse, takes us away from it.

HYLAS   How so?

PHILONOUS   First, it is just another way of your attempting to prove that continuity is not compatible with the parallel existence. Second, you make the concept of genesis identity essential for continuity. But it goes against the very principle of the machine's action. Consider. The tyrant orders his henchmen to cover your mouth and nose for long enough, and you die. A scientist who is a proponent of your genidentity doctrine, having carefully examined your corpse, declares that it is genidentical with Hylas and therefore Hylas's continuation, only not alive. He thus reveals a fact that is undoubtedly true but hardly novel, that a person, dying, turns into a corpse and that the deceased continues being the same person, only no longer alive. Such a revelation obviously brings nothing new to our issue whatsoever. You yourself discarded the genidentity proposition at the beginning of our discussion, when you said, correctly, that it is the preservation not of the atoms but of the structure that creates the perception of continued identity. Suppose the henchmen cut off your arms, and the machine creates new, living arms, which are attached to your body. Will you continue being yourself?

HYLAS   Of course.

PHILONOUS   And now the henchmen cut off your head, and I, using the machine, create a copy of your body, which is then attached to your head. Is it you who come back to life or your double?

HYLAS   It is me.

PHILONOUS   And if, after you have died, I create a copy of your entire body, head included, it will no longer be you?

HYLAS   Wait. A new idea just occurred to me. Before, you spoke about how we make observations, that is, about the methods that we select to determine whether the continuation of an object has taken place or not. Any such observation should be continuous, right?

PHILONOUS   Not in the least. Each of us, lying down after a long and difficult day, falls asleep and thereby loses the sense of our existence.

When you wake in the morning, notwithstanding this nocturnal intermission, you are perfectly aware of being the same Hylas as the one who went to sleep last night.

HYLAS   Of course! You are right. But listen, are we not giving too much weight to what goes through the mind of a man who is about to die? Perhaps the problem disappears when he does not know about his impending death. So he lies down, knowing nothing. When he falls asleep, we destroy him and place his atomic copy, also asleep, on the bed. When the copy wakes in the morning, can we not say that the continuity has been maintained, that it is the same person who went to sleep in the evening?

PHILONOUS   My dear Hylas, it has been a long time since I heard from you so many pronouncements spoiled by errors in reasoning. First of all, it must be unintentional (I do not wish to think otherwise), when you claim that if one kills a sleeping person or, in general, a person unaware that he is being murdered, we are committing less of a crime than if he were aware of his impending death. This issue, which belongs to ethics, I shall pass over in silence. Second, I am beginning to feel that entirely irrational and metaphysical fears are working in you. It is unclear why, when a copy is made after a person's death, you want it to be as close as possible to the place where the original died. As if putting the copy in the same bed, and asleep, created the ideal condition for the imaginary "transfer" of the "I" from one body to the other, from the one that has ceased to the one that has begun to exist. This is a manifestation of the irrational belief in the "I" as a sort of unitary, indivisible, and irreducible entity that has to be transferred between the two bodies, which is an example of the purest metaphysics one can imagine. It is not important that the appearances created by the external situation match our naive beliefs, for example, regarding the proximity of the deceased to the copy or the state of "being unaware" (perhaps you were thinking about the accident on the operation table and wanted to make the situation similar). The point is that we should be able to arrive logically at theses that apply to all circumstances in which atomic resurrection can be envisioned. How lame would be a theory of gravitation that applied only to apples falling to the ground and was useless in the case of pears or moons! Consider this following vision of the future. A person, embarking on a dangerous expedition to the stars, leaves at home his "atomic blueprint." When the news arrives that he has perished on the mission, his family turns on the machine, and out he jumps, alive and all smiles, to everyone's delight. If he has perished in the flames of Sirius, will you allow that his copy is his continuation, or does the distance between the place of death and the place of resurrection rule that out?

HYLAS   Surely, in both cases, mine and yours, there is no fundamental difference in the nature of the re-creation, so one must say that the copy is the person's continuation.

PHILONOUS   But what if the news turns out to be false, and the star traveler returns alive?

HYLAS   Then the family was in error, and the person created by the machine is a mere imitation, a copy, a double.

PHILONOUS   How then is the authenticity of continuation to be determined? Can it be the truth-value of the news about the traveler's death?

HYLAS   It can.

PHILONOUS   But what connection is there between information coming from the outer space and the structure of a man who is assembled atom by atom in the machine, and between that information and his thoughts and his whole personality? None whatsoever. Don't you agree?

HYLAS   You are right; there is no link.

PHILONOUS   Then how can something that has no connection with the person's identity decide whether the person is the same one who went to the stars or is only a copy of him leading a parallel existence?

HYLAS   I truly don't know. But let us tackle the problem in a different way. If you create the copy after the person's death, you can call it a continuation. The problem exists only in words; the fact remains that the person lives on. One might as easily argue whether or not yesterday's I continues to exist today. The difference between "the copy is the same person" and "the copy is the same as the person" is insignificant, as it does not change the fact of the actual existence. The dichotomy is therefore fallacious.

PHILONOUS   Fallacious? So there is no dichotomy at all? Really? One of two situations must obtain: either you perish in fifteen minutes from now by the tyrant's hand, and nothingness swallows you forever, while the produced copy, an exceedingly similar double, will perfectly compensate the loss for everyone except you, who will be dead—or thanks to the machine it is *you yourself* who open your eyes and see the sky and your friend, hear the birds sing, and feel the pleasant zephyr on your cheeks. Is there another, third possibility?

HYLAS   I don't know. Maybe there is. Let me think aloud. As long as a person lives, his continuation in a copy is impossible. Agreed?

PHILONOUS   Yes.

HYLAS   When he ceases to exist, his continuation becomes possible—for the rest of the world. That is certain. But for him . . . ? Grammar may be misleading us, because by asking whether continuation is possible "for him," we are talking about someone who no longer exists, which is the same as if he never existed, because his feelings, consciousness, and memory are no more. This is clearly a case of disallowed use of syntax.

PHILONOUS   Nice try! So now you blame syntax! Except that I am now speaking not with a dead person but with you a few minutes before you become a dead person. You mentioned earlier that we could not decide by words whether or not yesterday's I exists today. I see no difficulty making that decision. If I ask you where your robe of yesterday is, you understand me to mean the robe that you were wearing yesterday, yes?

HYLAS   Yes.

PHILONOUS   The robe is no more or less material than you; therefore, in the same objective sense, the yesterday's you also exists today. As for the subjective experiences that were experienced yesterday, they pose no problem either. Suppose your robe was creased yesterday when you sat on the porch of your house. A detailed examination of the robe reveals the displacement of some molecules in the fabric from yesterday's crease. This displacement can be called, metaphorically, the "memory" of the crease. So all objects, including our bodies, that existed yesterday continue to exist today. In contrast, our impressions yesterday—or, more generally, the states of our consciousness then—exist only in mind; their material traces are particular changes in the molecular structure of the brain that constitute memory. As you can see, there is no problem if we give the right meanings to our words. The statements "Yesterday's I exists" and "Yesterday's I does not exist" are both correct in the following sense: If by "yesterday's I" we mean my material, universally perceivable body, it exists today just as it did yesterday. But if by "yesterday's I" we mean the collection of thoughts and perceptions that manifested themselves in my consciousness yesterday, we cannot ascribe them an actual, current existence.

HYLAS   I admit my error. But how is this relevant to our problem?

PHILONOUS   It is not. Objectively a copy either is or is not a continuation of the original, depending on what convention we use to decide this. The whole problem stems from the fact that we are attempting to resolve the subjective aspect of the issue, that is, to determine by logic whether the mind of the deceased, his consciousness, which resided in his brain while he lived, reappears in the brain produced by atomic re-creation or this would lead to a contradiction.

HYLAS   Of course! All this time, we have been mixing objective with subjective. If our experiment cannot be carried out in an unbiased manner, logic is inapplicable, and our conclusions go out the window.

PHILONOUS   You think so? Very well then, Hylas, I will change the *modus operandi* to totally objective. The question of simultaneous versus sequential re-creation will disappear. And everything will be crystal clear and simple.

HYLAS   It will bring me great satisfaction. I am all ears, my friend.

PHILONOUS   No longer will I ask people about this and that or torment you, who are already suffering from the thought of dying, with pointless conundrums. I will simply kill you and then make copies of you—not just one, Hylas, but a legion of them. Thus, after you die (and you now have only five minutes left), you will exist as a collection of Hylases, Hylases beyond count, because I promise not to rest until all planets, suns, stars, moons, and all heavenly spheres and bodies are populated with them—such is my love for you. What do you say? To be everywhere in the universe, you alone!

HYLAS   It would be strange. Is there no logical contradiction in this?

PHILONOUS   That is for you to decide. These multitudes of Hylases live their lives, occupying themselves with various kinds of work and play. Here is the question: Does your "I," distributed among all of them, exist? Are the copies united into a whole by some mysterious unity of a single personhood?

HYLAS   No. Each must possess his own exclusive, private, subjective "I," except that it will be just like mine.

PHILONOUS   You are saying that each will have an "I" *like* yours? Then each "I" will not be your "I"?

HYLAS   It cannot be, because then we would all be the same person, which is a contradiction.

PHILONOUS   Excellent. Then each has an "I" just like yours, Hylas. But which one will have the same "I" and therefore be your continuation? Why are you silent? What does logic say?

HYLAS   Logic says, none of them. But wait. Something has dawned on me. Of course! This is how it is, my friend. Identity is determined not by the sameness of material but by the sameness of structure. We established that, already, didn't we?

PHILONOUS   We did.

HYLAS   And two structures can be either similar or identical. Suppose I draw an equilateral triangle. If I draw another, I can say that both share "the same" structural characteristic of equilaterality. I can draw many triangles like that, and from the structural point of view they will be in fact a single triangle repeated many times. Similarly, I can say that all those Hylases created by the machine are the same person, me, repeated many times. No?

PHILONOUS   Your conclusion is very clear and reasonable. So will you now allow me to make a copy of you while you are still alive?

HYLAS   What?

PHILONOUS   Since the copy, as you have just said, is, from the subjective point of view (and this is the one that matters), the same person as you, then it follows that when the tyrant kills you and the copy remains alive, then you too will remain alive, inasmuch as a person who is "the same Hylas" continues. Is it not so?

HYLAS   Could I have made a mistake somewhere? How is this possible? Does the same problem emerge when we consider inanimate objects . . . ?

PHILONOUS   It does, but we can dismiss it by shifting our point of view. Whether or not we accept a copy as a continuation of the original depends solely on how we agree to define a continuation. But in the case of a person, phenomenon of consciousness complicates the issue. We can easily mistake one earthen bowl for another when they look alike or one twin for the other when considered superficially as objects, but one twin will never mistake himself for the other. Likewise, you will never mistake yourself for a machine-made copy of you. What are we to do? The tyrant will come any moment now. Do you know the death he has planned for you?

HYLAS   First you said poison, then suffocation.

PHILONOUS   That was only for the sake of argument and did not coincide with his wishes. No, he plans to freeze your body until all movement, even the subtle vibration of your atoms, ceases, all tissues solidify and processes stop. Will it not be death, Hylas, when, in a block of ice, you are thrown into the maelstrom of the boreal ocean?

HYLAS   It surely will.

PHILONOUS   And when I, your faithful friend, pull that block from the abyss, thaw and warm up your body, and feed it and medicate it in such a way that all its molecules start moving again and you return to life, then what? Will it be you who stands before me among these autumn trees, having been freed from the icy prison, led from the darkness of nonbeing into the light of the day?

HYLAS   It will be me.

PHILONOUS   Beyond any doubt?

HYLAS   Beyond any doubt.

PHILONOUS   But when your atoms are scattered and I gather them and re-create you from them, it will not be you? Why not? Does your personal "I" fly away like a bird from the cage with broken bars?

HYLAS   In this case too, I now think it will be me who returns to life.

PHILONOUS   You yourself, Hylas, and not a man infinitely similar to you?

HYLAS   I myself.

PHILONOUS   Good. And when two copies of you are made, one from the same atoms that compose your body now and the other only from the same *kind* of atoms, then the first copy will be your continuation, your true self, whereas the second will be just your double?

HYLAS   I guess.

PHILONOUS   But the atoms of the same kind cannot be distinguished from each other. And the same applies to the two copies: they are indistinguishable. What then makes one of them your continuation but not the other?

HYLAS   I don't know. Indeed, there is no difference between an original atom and another atom of the same element.

PHILONOUS   Are both copies then your continuation? Or neither? Why are you silent? The seventh hour approaches, the tyrant will soon be here with his henchmen, but you, Hylas, even though I have presented you with all the possibilities of resurrection that a true materialist can envision, keep giving different arguments, now saying that your continuation depends on a prior agreement, now admitting that multiple continuations lead to an absurdity, now making your resurrection depend on the rate of decomposition of your corpse, and on and on. Give me your final word, my friend! I see the tyrant on the path, his robe stained with the ichor[3] of your predecessors. Tell me quickly what to do with the machine so that your resurrection will be secured and you convinced that it will be you who open your eyes after atomic reassembly.

HYLAS   To tell the truth, Philonous, I am at a loss. You have argued too well *per reductionem ad absurdum* that apart from the atoms and structures formed from them there is something else that makes a person's resurrection after death impossible, since the re-creation will not be the same person but only an infinitely close double. Could this be a proof, and you of all people its author, of the existence of a soul?

PHILONOUS Definitely not, my friend. All I have done is demonstrate the error of your thesis, which you have tacitly accepted as irrefutable, namely, that consciousness can be reduced to atoms or atomic structures. It is neither—*quod erat demonstrandum*. From which it does not follow, of course, that consciousness is not a material phenomenon. This issue is fundamentally complex beyond measure and it must be examined by other, new methods. Hopefully, we can progress from a criticism toward positive contributions when we utilize combined results of science in fields that appear remote from one another, such as psychology and the theory of electrical networks, or thermodynamics and logic. Only an investigation that makes use of the most recent achievements of science will enable us to advance the frontier of knowledge by another step.

II

PHILONOUS   Hello, Hylas. You're walking so fast through the park, I could barely catch up with you. Where were you yesterday? We were supposed to discuss that gem of knowledge that is cybernetics, remember?

HYLAS   You have no idea, my friend, into what confusion you threw me with your last argument. To make it worse, my philosopher friends are saying that your real purpose was to bring back irrationalism and undermine our faith in the cognitive power of the human mind and everything you said at the end (they say) was just to cover it up.

PHILONOUS   What am I hearing?

HYLAS   Really. Therefore, I decided not to share your conclusion with others, but to condemn it to oblivion instead. You must agree that your whole argument was completely negative. It only prohibited and sowed anxiety and doubts, offering no new, progressive idea.

PHILONOUS   Is that so? Well, my friend, let us think about it. But first, allow me to tell you a story. A long time ago, on a fertile plain, there lived a tribe in which some practiced hunting and herding, while others, fewer in number, strived to understand the world in which they lived, which is only human nature. One of these strivers, smarter than the rest, noticed one day that when he stood in the center of the plain, he saw only objects within two thousand steps; everything beyond that, whether it was a tree, a shack, or a person, disappeared so thoroughly as if it had never existed. He told others about it. They were unaware of this phenomenon, not having his sharpness of vision, but now, straining their eyes, they had to agree. Giving this some thought, they said to him: "Brother, you are right. However, your discovery may have dire consequences. It will raise a widespread impression that people and things that cross the limit of two thousand steps from our settlement get snatched away by dark forces, whereby it will promote believing in ghosts and similar dangerous superstitions. Let us therefore keep it from the public and forget it. You must agree that it only creates anxiety, sows uncertainty, and foments negativity, offering no new, positive idea and certainly not promoting progress." What do you think, Hylas, of this story? By the way, you surely guessed the real mechanism behind that discovered phenomenon . . .

HYLAS   Of course. We do not see distant objects because they are hidden from our view by the curvature of the Earth.

PHILONOUS   Correct. But the tribe did not know that Earth was round, and the first person who had the first hunch of it knew it, so to say, only in the form of a specific prohibition: that it is impossible to see distant objects.

HYLAS   You are saying, then, that your argument, in parallel with the story, also contains a rational, positive morsel of knowledge?

PHILONOUS   That's exactly what I'm saying.

HYLAS   Just convince me of this, and I will be the first to spread your argument to the world. What truth in it is equivalent to the curvature of the Earth in the parable?

PHILONOUS   Unfortunately, I don't know, just as the tribe's discoverer did not know. It often happens that the advancing human mind stumbles on a truth precisely by way of an uncertainty, a doubt, or an inability to do something.

HYLAS   So you have nothing to tell me?

PHILONOUS   But I do. Let me first recapitulate the argument. As you may remember, we were considering whether or not it is possible to resurrect a person by perfectly reconstructing his body from atoms and making a copy true to the original in every aspect. This assumption led to a contradiction, so we had to reject it. If I understand correctly, you want to know why this happened.

HYLAS   Yes. I also want to know whether or not a person's resurrection from atoms is possible and if not, why.

PHILONOUS   Then let us begin with that very question. First, we have to construct an exact map of all the atoms in the person's body, right?

HYLAS   That is obvious.

PHILONOUS   Obvious, yes, but is it doable? What does physics say? Heisenberg's uncertainty principle, fundamental to modern physics, says that we can locate an individual atom only in approximation: the atom's image is not a dot but a fuzzy spot, like an image on a photographic plate that moved during the exposure. For us, it is essential that the impossibility of precise location stems not from the technological inadequacy of the measuring apparatus but from a fundamental fact, a manifestation of the properties of the atom itself, which does not occupy space in the same way that macroscopic objects do in our everyday surroundings. If we cannot precisely pin down an atom, then we cannot draw an exact map of the atoms in an organism. From this the impossibility of creating an identical copy of a living person follows, *quod erat demonstrandum*. Are you now satisfied?

HYLAS   Not at all. Even if the uncertainty principle invalidates the goal of creating an exact copy of an organism, the difficulty is just technical, but we are considering the gnoseological, philosophical side of the issue.

PHILONOUS   Not so, Hylas. What you call "a technical difficulty" is in fact an integral, what's more, fundamental feature of the real world of atoms. Heisenberg's principle implies a specific "prohibition": it forbids the precise measurement of an atom. This "prohibition" is not an obstacle on our path to knowledge but instead an element of that knowledge. In the same way, the "prohibition" to see distant objects is no "technical difficulty" but instead a manifestation of a specific feature of the world, namely, that Earth is spherical. If the tribe's philosophers try to understand the geometry of their planet assuming that it is flat, they will inevitably come to contradictions and absurdities, just as we did. Heisenberg's principle does not exist as a kind of "prohibition" against making precise measurements at the atomic level. What we have inferred from countless observations and eventually formulated as a principle with that name is simply an inherent characteristic of the atoms themselves, which today we know only in the guise of this "impossibility." If atoms were not such complex objects manifesting traits as unique as the uncertainty principle but, say, tiny hard balls, our world would probably look completely different, and very likely it would not give rise to living systems or the neuronal structures that constitute the substrate of mental processes.

HYLAS   I don't understand. Are you telling me that the uncertainty principle is the very property of atoms that enables them to bond into systems such that life and consciousness emerge?

PHILONOUS   I wouldn't go that far. Look, it is the roundness of the Earth that makes the objects beyond the horizon disappear, true?

HYLAS   True.

PHILONOUS   The roundness also makes it possible for us to circumnavigate the globe and return to our point of departure, right?

HYLAS   Obviously.

PHILONOUS   But can we say that it is the disappearance of the objects beyond the horizon that makes it possible for us to circumnavigate the globe? No. Both these facts obtain because the Earth is a sphere, but there is no causal link between them. Now, the uncertainty principle is a manifestation of a specific property of atoms. Mental processes ultimately also derive from the properties of atoms. Only the future will tell what those fundamental properties are that enable both of these phenomena. Most probably the answer will not be as simple as in my story; rather, a whole chain of intermediate links, processes, and problems whose connection with what we discussed today no one even suspects will be drawn into the orbit of this issue.

HYLAS   For example?

PHILONOUS   That I do not know. I am not a prophet.

HYLAS   What you said reminds me of the often repeated claim that the uncertainty principle is a manifestation of the atom's "free will," from which allegedly the "free will" of human beings sprang.

PHILONOUS   The uncertainty principle does not represent an atom's "free will." This is a common misuse of language. I do imagine that there is a link between the properties of an atom and the emergence of consciousness, but I reject any such shallow and vulgar explanation of that link. The working of an electronic brain is indeed marked by the clarity of the reasoning process in the sense that the electronic brain functions correctly, logically, and without equivocation. In machines, such "thought" processes are represented by the flow of current through their electrical circuits. Lightning, which produces bright light, is an electrical phenomenon too. But to claim that the "brightness" of an electronic brain's reasoning derives from the "brightness" of the lightning would be utter nonsense—even though a link exists between the two, in electricity. A transfer of the uncertainty principle into mental processes in the manner that you mentioned is the same kind of nonsense. Such cheap and superficial analogies can be spawned *en masse*, and indeed are—by various metaphysically inclined know-it-alls, who just happen to practice atomistics instead of "mystics." But I digress. Have I convinced you of the impossibility of resurrection from atoms?

HYLAS   No.

PHILONOUS   Why not?

HYLAS   Perhaps Heisenberg's principle indeed rules out the atomic reconstruction of a dead organism. But it never entered our discussion, and therefore cannot possibly cause a contradiction in the course of our argument, don't you agree?

PHILONOUS   I do not. Dear Hylas, we did take the principle into account implicitly, when we improperly declared it invalid, and then it took revenge upon us.

HYLAS   I do not recall that it ever came up.

PHILONOUS   When we spoke about the copying process, we said that the atoms of the copy had to occupy the exact same positions as those in the body of the "original," right?

HYLAS   Yes.

PHILONOUS   *Atomos*, as you know, means "indivisible." Because atoms can be split (and with quite a dramatic effect), this term is now obsolete. Physicists might give atoms a new name, for example, "unlocalizables," that would better correspond with reality. So what did you and I do? We said, in effect, "The machine will place the unlocalizables into the copy. . . ." As you can see, we performed a self-contradictory, disallowed operation, which has no equivalent in the real world, at the very beginning of our argument. Is my explanation sufficient now?

HYLAS No. In my opinion the mystery and doubts that arose in our argument cannot all be attributed to that initial erroneous operation. It is well known that in specific circumstances we can create exact copies of atomic structures, for example, in the synthesis of simple protein molecules, which do not differ (structurally) from the originals. Perhaps those minuscule imprecisions of measurement, which the uncertainty principle warrants, are no obstacle in the creation of a copy of a living system. Nature, after all, can make copies of organisms that are incredibly similar—identical twins, for example. When people learn to do the same, they will encounter all the problems we have discussed here.

PHILONOUS Nature makes facsimiles, just as we do (e.g., the protein molecules you referred to). However, the absolutely identical localization of atoms makes something that is more than a facsimile, and the mystery of the emergence of mental processes in certain structures may lie precisely in this difference. But, as I said before, I am not entirely sure what truth resides in the core of my argument, and the issue of Heisenberg's relation, to which I pointed here, is just one possibility. There are others.

HYLAS I would love to hear about those other possibilities.

PHILONOUS Of one thing I am sure, Hylas: the "prohibition" on which we have stumbled, against the resurrection of a dead person from atoms, is a signal that we are using the notions of atoms and consciousness incorrectly, at variance with their true meaning. I already alerted you to the danger of applying the concept of atoms carelessly. It is possible that we have been equally careless about consciousness. A fundamental feature of consciousness is the subjective sense of its duration. Death causes a break in the thread of time, and once that thread is broken, it may not be possible to pick up the ends and reconnect them.

HYLAS Why? Isn't the thread interrupted in those who fall asleep or who temporarily die on the operating table (so-called clinical death)? We did mention those examples in our discussion.

PHILONOUS When going to bed, many people can decide in advance, and with success, when they will wake. So even as they sleep like a log, completely unconscious, the processes that count time must still be running in their brain. In clinical death too, the brain is active; electrical currents persist in the cortex and can be measured. So the fundamental brain processes continue in both cases and shut down and disintegrate with time only gradually, section after section. This breakdown is reversible up to a limit. When certain atomic structures in the brain suffer excessive damage and are disconnected, even the most basic processes break down and real death replaces clinical death. Only then can we say that subjective time has stopped. Perhaps its thread cannot be renewed for reasons that we do not know today, reasons that may be as fundamental as the one that precludes any body from reaching the speed of light. The latter reason was revealed in the theory of relativity, but consciousness is still awaiting its Einstein.

HYLAS   What you say smacks of sophistry. You are pulling down what you built with your own hands, without offering any new, positive ideas, which this *reductio ad absurdum* was supposed to contain.

PHILONOUS   I am not pulling down anything, Hylas, I am simply considering. Other factors may play a role too. Let me add just two. First, remember the question of whether or not the consciousness of the person reconstructed from atoms is the same as that of the deceased—in other words, whether or not it is the same person. We could answer this question only by interrogating and observing the person, but in principle one should strive for a solution that is objective, requiring nobody's testimony. For that purpose, one needs to be able to see and examine his consciousness directly, find a "direct link" to it, without the reliance on verbal pronouncements of the research subject.

HYLAS   But that is impossible.

PHILONOUS   How do you know?

HYLAS   Only the possessor of a consciousness has a "direct link" to it. Looking into a person's skull, you will see his brain but not his mind.

PHILONOUS   I am going to show certain indications suggesting that in the future we will be able to make these "direct links" into the consciousness of another person.

HYLAS Impossible! One cannot be oneself and simultaneously someone else, which linking directly to another's consciousness would mean.

PHILONOUS So you cannot be in one place and simultaneously in another?

HYLAS No.

PHILONOUS Not when you are projected by a camera?

HYLAS That's different.

PHILONOUS Well, let us see if I can convince you.

HYLAS I am all ears.

PHILONOUS First, you must tell me whether or not you accept my conclusion regarding resurrection.

HYLAS I do not.

PHILONOUS And why not?

HYLAS I still do not see the source of the contradiction and still do not know if it is possible or not to resurrect a person from atoms.

PHILONOUS What a disappointment! I've been telling you for an hour that I also don't know it for sure but I offered several possible solutions. And I have one more, which I have not mentioned yet. Suppose someone is dying, and two exact copies of him step out of the machine together. Which one is his continuation? It turns out that mere reasoning cannot answer this question, because reasoning follows formal logic, which disallows equations of the type $A = 2A$, labeling them "contradictory." But in our example, $A = 2A$ indeed seems to be the case, and a "multiplication of personality" has occurred. Perhaps one needs to apply a multivalued logic instead, which does not exclude the middle. At any rate, you can see that the impossibility of a solution could result not only from improper definitions of the starting terms ("atoms," "consciousness") but also from the employment of inappropriate reasoning tools (i.e., the logical system).

HYLAS I see now that you are not attacking rationalism. Yet . . .

PHILONOUS Yet what?

HYLAS I feel sorry for your previous argument. It sounded so convincing, clear, simple, but now turns out to have been a mistake.

PHILONOUS Not at all. It served its purpose—indicating the presence of something unknown. It pointed to hidden mysteries lurking where we thought there was certainty and everything was known. Is it not enough? In response, we need not close our eyes or sprinkle it with holy water and call it an argument in favor of irrationalism. Instead, we should address it, study it, and improve our knowledge about atoms and the processes of life and mind to the point where we will become able to tackle the resurrection problem with the objective tools. As you will see, cybernetics, or rather some of its consequences, opens the door to this.

HYLAS I have a feeling, my friend, that you might be the awaited Einstein of consciousness . . .

PHILONOUS Not at all, Hylas. There is a very long way from forming a question to finding the answer. I am just trying to formulate the question properly.

HYLAS And can you tell me what the question is?

PHILONOUS Sure, but not today. We will meet tomorrow. But my argument, is it now clear to you?

HYLAS No.

PHILONOUS Well, nor is it to me—but I think this is a good thing. A person who believes he knows and understands everything and sees no mystery is often on the path to peril.

III

PHILONOUS   What's the matter, Hylas? Where are you running to through these gardens?

HYLAS   Ah, I was just looking for you. I disproved your argument, Philonous. Don't be angry. It is so simple, I am surprised I didn't see it right away.

PHILONOUS   Which argument of mine did you disprove?

HYLAS   The one about atomic resurrection, your *reductio ad absurdum*. Everything you were saying about prohibitions providing a path to the discovery of hitherto unknown, new aspects of phenomena or new laws is totally irrelevant because your argument was false. It tricked the mind as a magician's trick deludes the eyes.

PHILONOUS   You don't say! Very exciting! So you disproved it, eh?

HYLAS   I did. I brought down your whole edifice of skepticism. Being a lover of truth, you will surely understand my intentions.

PHILONOUS   I don't take it personally. You will offend me only with poor logic or weak reasoning, never with its strength. So share with me what you found.

HYLAS   It was staring me in the face the whole time. We asserted that it is the identity of structure that solely determines the identity of a person, right?

PHILONOUS   Yes.

HYLAS   But structure independent of matter, or "structure as such," does not exist in the real world. Focusing on structure and neglecting the material aspect of identity, we came to the absurd. The absurd resulted not from a "prohibition" that pointed to the existence of hitherto unknown aspects of atoms or consciousness but from faulty reasoning.

PHILONOUS And where did the error in our reasoning lie?

HYLAS As I said, we disregarded the material level. If we modify your argument by saying that the identity of the physical particles of the human body joins the structural identity as a necessary condition, then the paradox of one person becoming two in parallel, that is, being simultaneously here and elsewhere, disappears.

PHILONOUS Are you saying that if we accept the identity of the structure as well as the identity of the material of the body as necessary conditions, then the paradox goes away and resurrection is possible?

HYLAS Without a doubt. If you construct a body that is identical to mine in both structure and atoms—that is, the atoms that constitute me now—then I come back to life. And the issue becomes utterly banal and deserving no further analysis.

PHILONOUS You are so sure of your conclusion, but let me tell you a story. Picture two people living on a desert island: you and your atomic copy.

HYLAS And? The copy is not me, because his atoms are of the same kind as mine but not the same as in my body.

PHILONOUS Correct. But bear with me. You are both castaways. There is nothing to eat on the island, and "the other Hylas," insistent, persuades you to agree to become his food. He devours you completely, bones and all, so that after a while, through metabolism, all the molecules that once made your body become his. I can therefore claim that you have become him not only in the structural but also in the material respect, because his muscles, bones, nerves, and brain now consist of precisely those molecules that previously constituted you. As you can see, the simultaneous identity of structure and matter, which you so much desired, has occurred. Yet you will probably say that you derive no benefit from this in terms of resurrection, because only your copy lives on, burdened by the awareness of being a cannibal. Or do you believe that your copy is now you, and it is you who walks on the island enjoying the ocean sunset? What will you say, Hylas?

HYLAS Only that you have defeated me again. I see clearly now that even when the identity of atoms is added to the identity of structure, such reconstruction does not enable resurrection. Yet, my friend, I can't accept the idea that our existence is transitory, so I seek, deep in the physical world, a gateway, a chance for us to reawaken after our death, a possibility to re-create the precious quality that is existence aware of itself. The delight of reasoning and perceiving should not be but a single brief flash, a tiny spark quickly extinguished in the dark of nonbeing. If the spinning wheel of material transformation is eternal, why can it not produce my thinking "I" once again, the "I" that is a living, feeling whole, irreducible to any of its parts?

PHILONOUS   What again? You think your "I" is a whole that cannot be reduced to its parts? You are definitely mistaken, my friend. It is, of course, an attribute of your living body; but at the same time, it is an abstraction, a generalization, and a resultant.

HYLAS   My "I" an abstraction? What are you saying?

PHILONOUS   If we followed the history of your "I" back to your childhood, we would come to a point when you did not refer to yourself as "I" but used the third person. All children do that between their second and third year of life. Talking and thinking of oneself as "I" requires the power of abstraction, which young children do not have. Hence an abstraction. Do you recall when you first became self-aware?

HYLAS   I do not.

PHILONOUS   You don't because it is impossible. You are you only because you have memory. Without memories, pleasant or sad, life events, worries, and hopes do not exist for you; you do not recognize your parents or friends; and you cannot learn even how to walk or see (as you surely know, newborn babies must learn how to see, that is, to assign shapes, meanings, and spatiotemporal structures to the colors and movements in their field of vision), talk, or think. You would be alive but blind, deaf, mute, and inert, and in no way would the awareness of being emerge in you. It is the memory of events, or rather their generalization, subordinate to the hierarchically highest centers of the psyche, that constitutes your "I." Hence a generalization and a resultant. A resultant of thousands of phenomena that involved you and in which you participated; of thousands of acts of selection, decision, and planning in response to what happened in the past; of the conflicts, defeats, and victories that affected both body and soul. Through summing all of this, day after day and year after year, your "I" took shape, until at last it assumed the designation of a mature person with full mental powers, that is, you, Hylas. But if we started erasing from your brain all the stored recollections, knowledge, and the automatic behaviors of walking, maintaining balance, seeing, and hearing, your mind would become poorer and poorer, until eventually you would no longer be you. Imperceptibly, you would become nobody, although your body continued to live. And the death of your "I," its annihilation, would be achieved when all the structural changes in your brain that took place in the past were destroyed. So your "I" can indeed be reduced to many constituents, and there is nothing strange about that.

HYLAS OK. I was so focused on the search for a repeated conscious existence that I overlooked such trivial truths.

PHILONOUS If you are so excited about it, we can have a closer look at the object of your search. Tell me more about it.

HYLAS I tried to be as rigorous as possible in my reasoning, and did come to a few modest conclusions. The question is this: Is it possible to calculate the probability in the future, after my death, of atoms joining and connecting to re-create the structure of my living body? Is not such a calculation similar to that of determining how many casts of n dice are needed to get all sixes?

PHILONOUS There is no such analogy as you have in mind between a series of dice casts and a series of Hylases existing in sequence. Could you be more specific about how you imagine a serial awakening of your consciousness?

HYLAS Last night I sat on the steps of my house and gazed at the stars. There was not a soul around. Facing the infinite, the stars, I felt that I was the only conscious being in the world, and it occurred to me that all the living beings—animals, plants, bacteria—that ever lived on the Earth from its beginning until now represent a microscopically small, negligible fraction of all the matter in the universe, which, in the terrifying vastness of nebulas and galaxies, is dead everywhere we look. How unlikely it is, I thought, for a lump of passive and inert matter, incapable of perception, to be pulled into the realm of life processes and turned into the tissues of a living entity, whereby it becomes the pinnacle of existence—a thinking creature. Yet this incredibly rare event took place precisely with the matter that makes up my body. This is what I was pondering under the starry sky. The matter filling my skull was once dead; for an eternity those phosphorus, carbon, oxygen, and iron atoms circulated in cold cosmic clouds until they found themselves on Earth; and after millions of years they were pulled into the orbit of evolution and ended up being my brain! If that happened once, why cannot such an accident happen again?

PHILONOUS   You should have become a poet, Hylas. The lyrical pathos of your words is moving, but I must say, with regret, that they have little cognitive value. I already showed you how your *post mortem* dust, your cold remains, get pulled again into the orbit of biological processes, as you put it, through your double's act of cannibalism on that desert island. Yet you conceded that this was no resurrection. Admittedly, the case was somewhat sordid in its details and far from your romantic yearnings under the starry sky, but we are concerned here not with the aesthetic value of phenomena but with their cognitive content, right?

HYLAS   You bring me down from the clouds again, my friend. Now sobered, I agree that I was expressing myself with insufficient care. But is it really impossible to calculate the mathematical chance of my renewed existence after death—in a purely materialistic sense, rejecting any metaphysical speculation regarding the soul or other similar fictions?

PHILONOUS   The question of your next existence is like asking whether yesterday's wind will blow tomorrow. You are an unrepeatable phenomenon, dear Hylas.

HYLAS   How so?

PHILONOUS   Forget the stars, the sense of solitude under them, the longing for immortality, and other lyrical things. The problem is simple. It is indeed possible to calculate the probability of the renewed emergence of a being structurally identical to you.

HYLAS   Aha.

PHILONOUS   Wait. Your body could arise, in some cooling nebula through an especially auspicious grouping of the same atoms as those that make up your body, but they all would have to meet and combine into the right organic molecules. Tell me, is it possible that in a bed of iron ore in a mine people discover a modern car formed by a chance ordering of the iron atoms into the chassis, the engine with pistons, transmission, wheels, and wires?

HYLAS   No, that's not possible.

PHILONOUS   Why? If you asked a physicist, he would say that his science does not rule out events that have low probability. The second law of thermodynamics, which claims that the most probable states occur most often, comes to the rescue. We can calculate that if people diligently comb through the iron ore deposits on 100 trillion planets of the galaxy for the next 70 quintillion years, they will find the car. But the spontaneous formation of a car from iron ore is far, far more likely than the spontaneous formation of a human body through a confluence of 80 trillion atoms. Let us say that the chance of such an event is one in a centillion. If the universe lasts forever, we have enough time, and this event, which appears to be so important to you, will eventually take place. What then? It turns out that you gain absolutely nothing from this "jackpot in the cosmic lottery," simply because "the next Hylas" has no connection to you, and there is no continuity between you and him. He may already exist as we speak in a quiet corner of the Canes Venatici nebula.[1] Yet it is he who is taking a pleasant stroll through its turbulences, not you, because every Hylas is a Hylas of his own. There is no memory link, no causal bond, between one Hylas and another.

That we have spent too much time on this banality is owing to your obstinacy in pursuing the phantom of the "atomic resurrection." The argument is mathematically sound, but talking about the universe in a hundred quintillion years from now is ridiculous and totally baseless. Please abandon these vacuous exercises of the mind in the attempts to grasp renewed life. There is another path.

HYLAS   Another path?

PHILONOUS   But it is long and arduous. To set out on it, we first need to consider many issues. Are you ready for that?

HYLAS   I am.

PHILONOUS   All right, then. I begin by drawing you a picture we will need later. We have said that the adult brain is shaped by storing everything that the person has experienced and learned in his life. All the memories, opinions, prejudices, knowledge, and skills are stored in specific transformations of its structure. Now imagine that just as one electrically charged molecule can transfer its charge to another, one brain can, through some kind of contact, transfer its full memory "charge" to another. Imagine that in this way two brains can totally exchange their structural features (which constitute individuality). Person *A* meets person *B*, this exchange takes place, then they separate, and all the personal characteristics of *A*—his temperament, talents, idiosyncrasies, habits, hobbies, addictions, and so on, along with the complete memory of all his experiences—now inhabit the brain of person *B*. The individual personality, we conclude, can jump from one physical body into another. Such a world contains no logical contradictions. We could even simulate it in real life by setting up a group of automata with brains designed so that during an encounter, one automaton would transfer to the other, through a series of electrical impulses, the complete pattern of its memory. Through this process, the first would become like a blank page of paper or a newborn baby and thus be ready to accept the charge of a different memory. As you notice, two features of human existence that in our world are inseparable are separate in this hypothetical world: the physical individuality and mental individuality of an organism. In our hypothetical world, a person, without changing his body, may occasionally become the carrier of a totally different mind. Curious romantic dramas could take place; for example, a Romeo might suffer from a predicament that to us is alien and incomprehensible—being unable to find the psyche of his Juliet after it took up residence in another body . . .

HYLAS   Now you are the one who is waxing poetic when we were supposed to address philosophical problems.

PHILONOUS   Sorry. I got carried away a little.

HYLAS   What is the outcome of this dramatic performance of yours in a cognitive sense?

PHILONOUS   We have come closer to understanding what personality really is or, more precisely, to understanding which physical aspects it can be reduced to. Namely, let us identify what it was that the brains in our example exchanged, what exactly was the phenomenon that we called the "memory charge." Well, it was the sum of the structural changes that the brain had undergone throughout its existence, in other words it was a set of certain pieces of information. With the word information we arrive at a crucial point in our thinking. It is information—its nature, origin, accumulation, storage, and utilization—that lies at the heart of cybernetics and at the same time is the key to the mysteries of systems like our brains.

HYLAS What is so special about information?

PHILONOUS It is something exceptional: it is a real thing yet it is neither energy nor matter. It cannot be a material object because the latter exists only in one place at one time, but the same information can be in many locations simultaneously (for example, in many copies of the same book). It can be measured with the methods of physics. It can be transmitted by material means. It needs a material substrate to exist but is not identical with that substrate. Matter is subject to a law of conservation: it cannot be destroyed, only transformed into energy,[2] but information can be irreversibly destroyed.

HYLAS So can it be another kind of energy?

PHILONOUS No, because as you are well aware, energy, too, cannot be destroyed. One kind of energy, for example, radiation, can be converted into another, for example, heat.

HYLAS Strange. What, then, is information?

PHILONOUS The importance of cybernetics lies in its answer to this question. Metaphorically, information is a child of thermodynamics turned upside down, as it is the opposite of entropy. Entropy is a physical measure of disorganization, disorder, or chaos in material systems. To explain the term without the use of mathematics, we must resort to examples and analogies.

In all processes that take place in nature, with no exception, in stars as well as in atoms, we observe an increase in disorder, an increase in the dispersal of energy.[3] A meteor possesses a certain internal order in that all its particles are moving in the same direction. When it falls into a bathtub filled with water, the ordered, unidirectional energy of its movement turns into a chaotic motion of particles, which manifests itself in the boiling of the water. We say that the ordered kinetic energy has been converted into the energy of disordered thermal motion. Fundamental in this phenomenon is its irreversibility. The opposite process—for the boiling bathtub water to suddenly cool and simultaneously cast the meteor back up into the sky—is impossible. Once condemned to the state of chaotic thermal motion, particles can no longer transform their energy back to the organized energy of the unidirectional motion of the meteor. The measure of this gradual energy disorganization and an increase in disorder, observed throughout nature, is entropy. A glass dropped on the floor breaks: the "ordered" energy of the fall descends to a lower level of organization. Something irreversible has taken place because the shards, left to themselves, will never join to form the glass again. The second law of thermodynamics generalizes phenomena like this and states that the entropy of an isolated system can spontaneously only increase but never decrease. This means that an increase in chaos and energy dispersal is the most probable and the most natural course, which is why only irreversible processes occur spontaneously. Gas under pressure in a container immediately expands and effuses when we open the valve, thereby decreasing the order in the energy of its particles. Warm bodies cool, because greater heat means a higher energy order in the system.[4] In nature the path leads from order and organization to chaos and disorganization. But information, to return to cybernetics from thermodynamics, is the opposite of entropy. It is the measure of order. If entropy points to the most probable outcome, information points to the less probable. In any closed system, information cannot spontaneously increase. It can only be destroyed, and once it is destroyed, we cannot re-create it in that system.

HYLAS   Why not? If we collect the necessary data again . . .

PHILONOUS   I said: in a closed system. It is a different matter if the system is in contact with its surroundings. Left to itself, cut off from any outside influence, whether it is a planet, mountain, or nebula, it exhibits an incessant increase in molecular disorder over time, a decay of structure, and the end of this process is the total disorder of energy and matter, a whirlpool of completely randomized atoms.

Eroded rocks do not rise from the scree and solidify again on their own, meteors do not shoot back to the stars from which they came, and a broken crystal cannot fuse without an input of external energy (e.g., from the sun). The spontaneous increase in entropy has been confirmed in nebulas and stars, in the heavens and on Earth, and yet there exist systems that seem exempt from this universal rule.

HYLAS   You mean our bodies?

PHILONOUS   Exactly. The fertilized egg cell (a zygote) has a lower organization than the adult organism that develops from it. It seems as if the phenomena of life went "against the stream" of all natural processes. Beyond the sphere of the living systems, we observe only an increase in disorder, decay, destruction, and the simplification of structures, whereas the entire course of biological evolution is the opposite, an incessant decrease in entropy, a progression in complexity from parent forms to descendant forms.

HYLAS   But life does not violate the second law of thermodynamics, Philonous. We know that living systems are not closed; on the contrary, they live thanks precisely to their surroundings, as they grow and develop at the cost of the consumed meals, whose organization is lowered in the process of digestion. Animals live on plants, and plants in turn support the synthesis of their tissues by using the radiant energy of the sun, which thereby undergoes disorganization. The overall thermodynamic balance still shows an increase in entropy.

PHILONOUS   You are right, except that this overall balance, which confirms the validity of the second law of thermodynamics, does not in the least explain the phenomenon of life. Consider: a device or machine can only make something that is (structurally) simpler than itself. A machine for making shoes is more complex than a shoe and a machine that makes nails is more complex than a nail.

HYLAS   Must the maker always be more complex? Cannot the maker and the made be equal? Take, for example, a casting machine and what emerges from its die.

PHILONOUS   The machine is always more complex, my friend.

HYLAS   Wait. A lathe, which is a relatively simple tool, can produce a very complicated object.

PHILONOUS Only if a person runs the lathe, and then you have the machine plus the human brain with a structure whose complexity has no match in the universe.

HYLAS How about a computer calculating an extremely difficult problem? Cannot the assignment be structurally more complex than the machine used to solve it? Not that I have any idea how to measure differences in complexity . . .

PHILONOUS The complexity or "complicatedness" of a structure is, in our cybernetic approach, simply the amount of information it contains. A computer can perform a task that is structurally more complex than itself only if we provide it with appropriate instruction. But the instruction also is a kind of structure or, more precisely, information. So the computer plus the instruction outweighs the product—the solution—in complexity. To explain, let us take your casting machine. It produces, say, human masks from a mold, that is, the mold transmits to the clay a specific amount of information. But in practice the mold's tiniest details gradually erode, and each next mask will be less and less complex or detailed (poorer in information) than the original mold. This is a manifestation of a universal principle that says that in a transmission process, information decreases or loses quality, but can never spontaneously increase. As you see, this is a "reverse," cybernetic version of the second law of thermodynamics, which rules out a spontaneous increase in entropy. But let us return to our mask casting. If we proceed with casting a new mask from a mold produced from the previous mask, eventually we will end up with a mask that bears little resemblance to the original and contains no facial subtleties. Note that this degenerative tendency is not present in the reproduction of living organisms. Otherwise children would always be organizationally poorer than their parents, and after a certain number of generations their systems would become too disorganized to support life.

HYLAS So the rules of information transfer do not apply to living creatures? Or it is not only the information contained in the egg cell that creates the organism that develops from it?

PHILONOUS The laws of the circulation and transfer of information, just like the laws of thermodynamics, hold universally, in all systems, whether living or not. But something special happens in evolution that prevents the degenerative tendency from manifesting itself. This something, absent in inanimate nature, is the crossing of the minimum complexity threshold. What does this threshold mean? All systems below it are incapable of producing other systems whose complexity matches theirs. Above it, the creative apparatus can produce systems as complex as itself.

HYLAS   But wait. In evolution, simpler organisms give rise to more complex ones. So in certain circumstances, less information may give rise to more. If that is true, then the cybernetic law saying that information cannot increase during transfer does not apply to evolution. What do you say to that?

PHILONOUS   The cybernetic law is not violated by evolution. An organism, when producing offspring more complex than itself, does not "create" the information out of nothing but draws it from the environment, just as it takes food from its surroundings to prevent the increase of internal entropy, the food's energy disorganization through metabolism balancing its thermodynamic bill. How does the organism draw information from the environment? There are two ways. First, its nervous system takes in information, a gain for the organism but not for its offspring, because, as you know, an individual's memories of experiences are not transmitted genetically. Second, information is drawn from outside by the process of reproduction and the succession of generations.

HYLAS   I don't understand. Can you explain?

PHILONOUS   We need to introduce here the second fundamental concept of cybernetics, namely, feedback. Feedback means that information about how an organism's actions affect its surroundings is fed back to the organism so that it can modify its future behavior. It is the mechanism whereby a system becomes self-regulating because it can continually correct its subsequent actions to attain a goal. When I put out my hand to pick this leaf from the ground, information about the effects of my movement runs by feedback through my eyes to my brain, so that if I am reaching too far or not far enough, the visual image lets me know, and I can immediately make the appropriate muscular adjustments.

HYLAS   That is clear.

PHILONOUS   Feedback operates in evolution as well, except that it goes not to the same organism but to the next generation. The organism "acts on the surroundings" by producing offspring. If that act was "on target" (like my success in picking up the leaf), then the offspring will survive in the world and reproduce, providing future generations. But if the act is "not on target" (my hand missing the leaf), then the environment, acting as a filter, does not let the offspring pass through and "introduces a correction"—the offspring does not survive. Hence feedback in evolution operates through the cycle of generations. The evolutionary adaptation of organisms is equivalent to a change in the information contained in the reproductive cells. This information increases via genetic mutation. Most mutations do not add useful information and therefore are eliminated by the environmental filter. Only those that prove useful, that increase the ability to survive, pass through.

It is a very slow process: a single feedback-regulated act of an individual is in evolution equivalent to the lifetime of an entire generation. But evolution has time—on the order of two billion years. What is the source of the new information that accumulates generation after generation in the chromosomes of the reproductive cells of the organisms that pass through the filter? The information increases at the expense of the disorganization of the photons from the sun, which ultimately enables the existence and development of life on Earth. Is this cybernetic interpretation of the evolutionary process clear to you now?

HYLAS   Not at all. I particularly dislike the idea that the reproductive cells (gametes) gain information thanks to the random, blind hits that are mutations. I read somewhere that the amount of information in the gametes' genes is roughly equivalent to the amount of information contained in the *Encyclopedia Britannica*. How likely is it that those forty thick tomes could be printed by tossing the pieces of type on the floor, reading the random result, and removing the meaningless combinations of letters? It seems to me that even if we repeated this process for billions of years, we would not end up with the encyclopedia. This is how I view the cybernetic picture of evolution. Also, we should be able to calculate the probability of that information accumulating in the human gamete after two billion years of evolutionary feedback operating in the system of blind mutations and the environment's selection filter. If the probability turns out to be negligible, which I expect, then we will have to conclude that acquired traits are heritable, no?

PHILONOUS   The matter is more complicated than that, Hylas. In the statistical sense and also in terms of the amount of information there indeed exists a similarity between an encyclopedia and the gamete, as both are carriers of information. But in the dynamics and internal rules of their systems they cannot be compared. Printing an encyclopedia from random pieces of type has nothing to do with the operation of the evolutionary feedback links that determine the mutation genotype distribution of a population.

HYLAS   Why not? Doesn't population genetics use the same statistics that gives the probability of assembling an encyclopedia by tossing pieces of type on the floor?

PHILONOUS   Except that the "selection filter" of the environment acts only on organisms that have been already born. Therefore, a new factor enters the picture: the phase of embryonic development, that is, the dynamic expression of the information in the genes, the process that turns the given store of biological structure into a living offspring. An encyclopedia is a catalog of information packaged side by side, and no entry in that catalog has any effect on any other entry. Whereas in the cell, a change in the information (a gene mutation) often has far-reaching consequences on the development of the *entire* embryo. It is precisely because of the internal cohesion and interaction among the elements that the gamete and an encyclopedia are not equivalent and therefore cannot be studied with the same methods.

HYLAS   What do you mean by "internal cohesion of information" in the gamete? That the information is "dedicated," as it were, to the task of creating or constructing an organism, while the information in the encyclopedia has no such unidirectional purpose? In that case, let us replace the encyclopedia with a thick volume describing, for example, how to build an atomic power plant. The analogy then will work, as in both cases we have information dedicated to a *single* issue.

PHILONOUS   Not at all. The manual will not build the atomic power plant by itself, whereas the zygote makes the descendant organism on its own. The zygote is much more than merely a "blueprint for construction"; it is at the same time a set of feedback links enacting the actual process of construction. In the embryo, thanks to internal feedback, the circulation of information regulates its development. If you built an electronic brain capable of transforming instructions into actions but tore the last twenty pages from the manual, the brain would not have sufficient information to succeed. But an embryo, even when damaged (provided the damage is not too extensive), can compensate for the missing information and still produce a healthy, normal offspring.

HYLAS Why can the embryo compensate for the damage when the electronic brain is unable to fill the gap in the instructions?

PHILONOUS One could construct an electronic brain that would have the ability to fill a gap in its instructions autonomously (e.g., it could complete the data with results from its own, independently conducted experiments). But such a machine would be far more complex than one that does no more than faithfully and blindly follow instructions. The difference between the two is the capacity to learn. Because the zygote corresponds to this more complex brain, we arrive at the surprising conclusion that the zygote can learn. Which is indeed the case, as the mature organism represents a structure far richer in information than the embryo. The embryo accumulates the new information during its development thanks to the internal feedback links. Mutual adaptations of forms, chemistries, and functions continually enrich it in information.

HYLAS This is a little hard to believe. An embryo can learn? And enrich itself in information? How does that happen?

PHILONOUS It happens thanks to the universal ability to react, which every living tissue, including (or, perhaps, above all) embryonic tissue, exhibits, in conjunction with the interconnectedness or integration of this reactivity through system-wide feedback. All tissues and organs learn their functions in the course of their development. The heart of the embryo has barely formed but already is beating; as the blood of the embryo circulates, it strengthens the walls of the vessels. In a word, the set of genes in a chromosome activates gradients of development instead of imposing a strict timetable; chemical reactions influence each other to shape the organs, cells, and tissues while they are already functioning. Hence the zygote is "a set of building instructions that is able to learn," that is, to absorb additional information that regulates its development. A mutation is a change in that set that can affect the entire process of the construction work, not just one trait. That is one reason why it is so difficult to analyze the whole phenomenon mathematically.

HYLAS   You said nothing about the zygote crossing the threshold of the minimum complexity. Could that be another difference between an encyclopedia and the reproductive cell?

PHILONOUS   Absolutely. The hypothesis of the threshold of complexity explains many things. It explains, first, why living beings are systems so exceedingly complex and why there cannot be a living organism with a complexity on the order of that of a conventional machine or mechanism. It is because a simpler structure, subject to degenerative tendencies, would die out after a few generations. Second, the threshold of the minimum complexity represents a well-defined, physically measurable border between the world of mechanisms in the classical sense (machines) and the world of *organisms*. Note that I do not say "the world of *living* organisms." Here, "life" is a narrower term; "organization" is broader, better. This new division suggests the possibility that there may be organisms or systems constructed from inanimate elements or parts that behave like the living systems made of proteins. A "nonliving organism," in this sense, does not denote a corpse but rather a system made of some inorganic material, such as glass, silver, or nickel, but so complex that it has passed the threshold and acquired the ability to self-reproduce, self-repair, store, and utilize information collected during its existence, and, finally, strive to accomplish goals. Do you follow?

HYLAS   I don't really see the point of a nonliving organism. What does it have to do with our discussion?

PHILONOUS   I will try to explain that—but not today. In the time that separates us from our next meeting, please familiarize yourself with the fundamental notions of information, entropy, and the complexity threshold, because it is on these pillars that we will erect the magnificent edifice of cybernetics.

IV

PHILONOUS   Hello, my friend. Why so gloomy, though you sit by such a beautiful brook?

HYLAS   Hello, Philonous. Indeed, I confess that neither the beauty of the landscape nor anything else pleases me after you shattered my most firmly held beliefs. I feel I know nothing anymore. The mystery of consciousness, just thinking about it depresses me: in a thousand years, we have not advanced a single step in understanding it. The sciences approach perfection and terminology cannot be more subtle, yet the same chasm yawns between consciousness and matter, and any attempts at analysis end up in an awful *circulus vitiosus*, in which the mind is trapped as if on a treadmill with no exit. Abomination!

PHILONOUS   Too many words, so much confusion and chaos! What are you saying, Hylas? What chasm? What mystery of consciousness?

HYLAS   If I examine a person as a neurologist would, I will find that sound waves reach the ear and transform into nerve impulses that travel to the brain, where they are sent to a part of the cerebral cortex. From there the impulses travel through nerves to the muscles in the hand that the person raises. I can describe the entire process, from the arrival to the ear of the sound waves carrying the command to raise the hand to the command's execution as a chain of physical causes and effects, in every consecutive phase of which atoms participate—nothing but vibrating, dancing atoms. There is no place in this chain where consciousness might be hiding—unless a certain subset of those dancing atoms is consciousness. But how is it possible that one group of atoms is, and another group is not, consciousness? What are atoms but a vacuum in which minuscule electrical charges circle in quantized orbits, with their waves of probability, spins, magnetic moments, and the devil knows what else? Does consciousness then consist of a vacuum and electrical charge? I don't know anything anymore. When in another experiment I take the place of the examined subject, it is my ear that receives the sound waves and my hand that receives the command to raise, and I am fully conscious of the entire process. In the first case, we examined a physical aspect of the process that can be observed externally; in the second case, we experienced its mental aspect. The physical aspect is accessible to everyone, and this kind of phenomena is therefore called public; but the mental aspect is directly accessible only to me—no one else can know whether I raised my hand intentionally, with consciousness involved, or automatically, due to a conditional reflex.

PHILONOUS None of this is news, Hylas. Why does it torment you?

HYLAS How can you ask? Some say that physiological phenomena do not affect the mental ones and *vice versa*, that they both run in parallel, mutually reflecting each other. This is the view of parallelists. Others, epiphenomenalists, claim that mental phenomena merely supplement or passively reflect the physiological ones and are perceived through the subject's "inner sense." There is also a theory of "two sides," according to which the physiological and the mental are two aspects of the same thing: when I look at a process from the outside (e.g., when I observe your brain), I see it as physiological, but when you experience it "from inside," it appears to you as mental. Then there are spiritualists and materialists of all stripes and, to top it off, physicalists who sincerely inform us that the entire problem of consciousness is bogus, wherefore scientists can say nothing about it. As if we can walk but cannot even mention our legs. I begin to suspect that a principial "impossibility" of knowing the actual state of the matter is at play here. Is there really no transition from what happens in the mind to what everybody can observe? Could this "prohibition" hide within itself a fundamental mystery of nature just like Heisenberg's "prohibition" of a precise observation of an atom?

PHILONOUS Don't be so quick to posit a "prohibition" and despair. What do you think about consciousness?

HYLAS If I only knew! First, it is not a thing or an object but a process or a series of events. Second, it manifests itself exclusively in a living organism at a high stage of evolution, such as a human being. Third—

PHILONOUS   Only a living organism can be conscious?

HYLAS   Surely, Philonous, you do not doubt this point.

PHILONOUS   I do not doubt that consciousness can arise in any system that has certain characteristics, but life is not among them.

HYLAS   Are you suggesting that consciousness could arise in a structure built of, say, glass and metal?

PHILONOUS   Exactly.

HYLAS   But you are contradicting the obvious.

PHILONOUS   Please, Hylas, refrain from using the word "obvious" in our discussions. Many rifts in philosophy occur precisely because what is obvious to one is not obvious to another. It was not obvious once that people didn't walk upside down on the opposite hemisphere or that simultaneity cannot exist between events occurring on two stars far away from each other. For me, a scientist, nothing is entirely "obvious" or "self-evident"; every statement deserves a rigorous analysis and an experimental confirmation of its consequences. So do you maintain, Hylas, that consciousness cannot arise in a structure made of metal?

HYLAS   I do. Can you prove otherwise?

PHILONOUS   I will show you a place where metal is the source of consciousness.

HYLAS   Indeed. And where is that?

PHILONOUS   Here, in your head.

HYLAS   You are joking.

PHILONOUS   Not at all. You must know that your body and therefore also your brain contain iron—in the molecules of respiratory enzymes. Without those enzymes, hence without the iron in them, you would not think or even live for a second.

HYLAS   True, but—

PHILONOUS   Because iron is an inseparable element of tissues including those that form your brain, it plays a role in the processes that are the basis of consciousness. *Quod erat demonstrandum.* What do you say to that?

HYLAS I say that when iron is in an organic molecule, a protein, it loses its regular properties—

PHILONOUS I must interrupt, because you are talking nonsense. What properties does iron lose when it becomes a part of the enzyme oxidase? Perhaps those that show in a nail or a horseshoe? So you think the iron in a horseshoe is ordinary and that in an enzyme not?

HYLAS I only mean that this iron—as a part of a greater whole—has been drawn into a system of life processes.

PHILONOUS Whereas in the horseshoe it appears in its perfectly pure form, isolated and autonomous? Is there any "immanent" iron? If we agreed to such a terminology, surely the criterion of purity of the element would not be met by a horseshoe, whose microcrystalline structure is full of various additives. Is it not better to say that the iron atoms found in a crystalline structure with additives of carbon and sulfur exhibit properties that we observe in a horseshoe, whereas the iron atoms complexed with a protein molecule exhibit properties that are different? In each case, we observe a manifestation of particular characteristics in a particular system, that's all.

HYLAS Are you saying that the presence of the iron atoms in my head implies that if my head were made exclusively of iron, I could think, feel, and have consciousness?

PHILONOUS Provoked, I must respond: yes, this is exactly what I am saying, but with one important reservation—that the iron is in a structure functionally equivalent to your brain.

HYLAS You are being careful, Philonous, but is it not just playing with words? You say "an iron brain would think if it had the same functional properties as the living brain" or "an iron brain would be alive if it had the same functional properties as the living brain." But an iron brain cannot be alive; therefore it cannot develop consciousness. An iron brain is a *contradictio in adiecto*.

PHILONOUS I think that this way, we get nowhere. You maintain that the processes of life and consciousness are inseparable and specifically, that not every life process is conscious, but every conscious process is also a life process. Yes?

HYLAS That is my opinion.

PHILONOUS Then we must first consider what exactly consciousness is. We, incessantly talking about it, find this entity as essential for survival as, e.g., sight. But in reality, consciousness is an abstraction. When I say that I am conscious of something, it means that I understand it, perceive it, or think of that something. Nothing more. Because when I see or think, it does not mean that I also, somehow "in addition" or "above and beyond" that perceiving or thinking, have awareness of it. When you are conscious of me approaching, it simply means that you see me coming closer, nothing more. Do you agree?

HYLAS In principle, yes. Normally, I do not see an object in such a way that I am conscious of being conscious of seeing it, but when I deliberately concentrate on that act of seeing, it does happen that, in addition to the seeing itself, I become aware of the act of seeing.

PHILONOUS Of course we can make ourselves aware of an act of perceiving, but so what? Seeing me, you may think, "I see Philonous" or "I am conscious of seeing Philonous," where your awareness of seeing coincides with your seeing. Just as I can merely sing, or sing about my singing (that is, I sing that I am singing), I can say to myself that I perceive that I am perceiving. I am attempting, when I do this, to generalize the experience while it is happening, that is, to assign it to the class of phenomena that are called conscious. You likewise are attempting to create a generalizing distance from the simple act of seeing, which sets you on the path to the creating of the abstraction "consciousness in general." If I think about something, I simply think about it, but I can also think about the thinking process itself. If "thinking about thinking" is qualitatively different from "thinking about eating" and you accept the idea of "consciousness of consciousness," you also must accept the possibility of a next level, to wit, Hylas is thinking about thinking about his thinking, which gives you "consciousness to the third power," and this *regressus* may be repeated *ad infinitum*. The inevitable conclusion is that there are infinitely many levels of consciousness. Which is absurd. Obviously we can abstract on any topic, including the topic of abstracting, but each such act of the mind must be limited to a single, defined topic. The idea of consciousness, as you can see, comprises diverse mental phenomena—feeling, thinking, seeing, and so on. If you ask a child of six or seven if she has consciousness, she will have no idea what you are talking about, yet you would not deny her consciousness, would you?

HYLAS So in your opinion consciousness does not exist at all. You have cleverly "explained it away," and nothing remains to be said on the subject. Are you suggesting that the problem is just apparent? Have you become a physicalist? As for the example of the child, don't forget that one can speak in prose without having any idea what prose is. One can be conscious without being conscious of it.

PHILONOUS A good argument against those who believe that "Hylas's consciousness" is qualitatively different from "Hylas's existence," that consciousness, instead of being a generalization of a series of phenomena, is a primal property, fundamental, obvious, and absolutely *a priori*. Only philosophers, thanks to their habituation to this notion and being "professionally conscious," as it were, came to this conclusion. "Speaking in prose," and even "prose" itself, is a generalization, just like consciousness, and we can live just fine without making those generalizations. When I say, "This is prose," I am actually saying, "If we take into account certain stylistic, rhythmic, and other features, this articulation can be characterized as prose and therefore belongs in the category of all possible sentences expressed in prose." Similarly, consciousness is definitely not an "apparent problem" but rather a generalization of a series of mental phenomena that we group into a category under the name "the phenomena of consciousness" or, in short, consciousness. With this in mind, let us now build "your iron head that thinks." We will put together an electronic brain made of iron (or another metal). Is this possible?

HYLAS It is. But that brain is dead matter.

PHILONOUS It is dead matter, and yet it is capable of thought.

HYLAS Thought that is only formal, that only applies certain operational rules to certain signs (symbols). Such a brain cannot think meritoriously. Even some cyberneticists refer to the processes taking place in an electronic brain as "pseudo-thinking."

PHILONOUS True. If you like, we can use the term "pseudo-thinking," even though we will use its results exactly as we use the results provided by living calculators. We can also reserve the name penicillin only for the antibiotic produced by a living mold and call synthetic penicillin "pseudo-penicillin." Why not? I just don't understand what we want to accomplish by making that distinction. Is the intention to put up a wall between the electronic network of the electronic brain and the neural network of the protein-based brain? What is the worth of such a wall if it is only made of words? Why limit ourselves voluntarily by saying up front that this or that is impossible? Would it not be better to examine the question of impossibility *sine ira*, with the methods of logic and empiricism?

HYLAS   Very well, I withdraw the word "pseudo-thinking." But I still retain the conviction that no electronic brain can think substantively, that is, with comprehension and subjective understanding.

PHILONOUS   First, you will have to prove that consciousness cannot arise where there is no substantive thinking. But let us proceed step by step. We construct our electronic brain and give it powerful "word memory stores" and a scanning device so it can read. Is this possible?

HYLAS   It is.

PHILONOUS   Can we have a conversation with such a brain?

HYLAS   How?

PHILONOUS   Of course, we would need to attach to it a device for analyzing sound waves. Such a device already exists as a prosthesis for the deaf. We would also have to scale up the brain considerably, making it much larger than any in existence today.

HYLAS   Yes, this is possible.

PHILONOUS   Our brain now possesses an organ through which stimuli from the surroundings enter it, as well as an organ through which it sends impulses to its surroundings. Input and output. Now we can communicate with our brain. Agreed?

HYLAS   We can communicate with it only in the sense that if we assign it a certain task, it will solve it (provided it can). But the reasoning with which it performs each task will be purely formal and not substantive.

PHILONOUS   Your caution is exemplary. Therefore, let us open up the parentheses here and consider what substantive, as opposed to formal, reasoning is. My guess is that you understand it in this way: when I say that a straight railroad track, when observed along its direction, gives the impression that the two rails meet at the horizon, you will understand me at once, right?

HYLAS   I will.

PHILONOUS   So you grasp my meaning without resorting to formal methods, without using the laws of geometry and the canons of physiological optics. How can you do this? Because you know or "intuitively feel" that the rails in fact appear to meet at the horizon. Can our artificial brain do the same?

HYLAS   Yes, but it would test the truth-value of your statement only by formal reasoning. We would need to provide it with definitions of terms like "rails," "the horizon," and "meet," as well as with instructions on the operations to be performed on them using the laws of geometry and optics that you mentioned. Then and only then could it arrive at the correct conclusion.

PHILONOUS   Excellent. Now imagine a person who is paralyzed from birth, blind, deaf, and mute, and has no sense of touch except on the palm of one hand. With great effort we have taught this person about his surroundings, by drawing letters on that palm. Now I tell this poor creature, letter by letter, that when one looks at a railroad track, the rails seem to meet at the horizon. Would he understand me at once? Would he immediately grasp the substantive meaning of that phenomenon?

HYLAS   ...

PHILONOUS   You are silent, realizing that the man, although alive and possessing consciousness (his brain functions properly), cannot comprehend the issue because he lacks personal experience with such terms as "to look," "a distant object," "a near object," "optical perspective," and so on. Yet he can understand what I am telling him. How? By using, yes, formal reasoning. We can teach him geometry and optics (since the laws of both can be expressed in the formalized language of mathematics), and by applying those laws to the problem, that is, through formal reasoning, he can integrate the statement "the rails seem to meet at the horizon" into a logical whole and declare it to be true. As you can see, what for some may be a matter of formal reasoning only, for others may be grasped directly, without resorting to roundabout ways. Please realize that the only instrument of sensation in an electronic brain is the analyzer that reads the perforations on an input card or tape, and for this brain the whole external world is reduced to those perforations. This link with the world is thus even more tenuous than the disabled person had. For an electronic brain to be able to think substantively, we would need, first, to augment enormously its circuitry to provide it with vast possibilities for forming links (associations) between impulses, and, second, to equip it with organs for many kinds of contact with the external world—optical, tactile, chemical, and so on.

HYLAS   Why are the engineers not working on this now?

PHILONOUS   The engineers are interested not in aping human behavior but in building instruments that efficiently perform narrowly specific tasks. Our current electronic brains are "idiot calculators," combining the highest speed and precision of formal-mathematical reasoning with great stupidity in all other areas of mental work.

HYLAS   Then you believe that, if given the sensors and circuitry needed, an electronic brain would be able to think substantively?

PHILONOUS   I do. But I am not belittling the obstacles that lie in the path to the construction of such a brain. We can talk later about the prospect of that construction. My point here is different. Suppose that we have built such a brain. You come to see me, its constructor, and you see the machine reading a book. You ask the machine what it is doing. I am reading, it replies. What are you reading? I am reading a book, it says. And who is reading the book? I am, answers the electronic brain. So it has an "I," it can read, it can see, and, if equipped with the right circuits if you offended it, it would say that it is offended. Therefore, it also has feelings. Since we have agreed that to feel, read, and perceive is the same as to have consciousness, our augmented brain will be conscious, *quod erat demonstrandum*. What do you say to that?

HYLAS   That consciousness cannot arise where there is no life.

PHILONOUS   How do you know this? So far it has not happened because there have been no electronic brains. But now there are.

Granted, an electronic brain with which one could communicate as we have described does not exist yet. It would need to be a million times more complex than the machines today, but that is a technical question, which for us, theorists of knowledge, is beside the point.

HYLAS   There must be an error somewhere in your reasoning. Why are people—all organisms, for that matter—not made of iron, nickel, or glass? Why are there no inanimate minds? Why was there only one evolution—the biological one, I mean—and why only it has managed to produce beings that are immeasurably complex? Does this not prove that the increase in organization and life are inseparable, that neither can appear in nature independently, just as there cannot be matter without mass?

PHILONOUS   Finally, you are on the right track. Let us take a closer look at this issue. You ask why we are not made of metal and glass but of colloidal proteinaceous compounds instead. I will try to answer. First, although the organs of our body are indeed made of living tissue, living tissue is not necessary for them to function properly.

HYLAS   What do you mean?

PHILONOUS   Take, for instance, a heart, a blood vessel, or a kidney. An artificial, mechanical replacement of any of them functions well for an extended period.

HYLAS   That is true.

PHILONOUS   Let us then define certain sets. All possible systems that perform the same function will belong to one set. Only the function decides; the building material, size, and technical or structural details are irrelevant. Thus the set of all possible pumps will contain piston and pistonless pumps, centrifugal and vacuum pumps, absorption and mercury pumps, and so on. In this set there will also be the hearts of living creatures. In the set of all possible filters there will also be the kidneys of living creatures. And in the set of feedback networks there will also be the brains of living creatures.

HYLAS   How is this an answer to my question?

PHILONOUS   It is just the preamble. We have established that the function of some organs in our body can be continued, even improved, by devices made of nonbiological material. The field of prosthetics is developing slowly but has great potential. Cybernetics has contributed to this field significantly. Advances are being made today for prosthetics for the deaf and blind. But let us return to your question of why we are built out of colloidal proteins rather than metallic conductors, tiny wheels, screws, and so on. When, as a constructor, I set out to design a heart or kidney prosthetic, or an artificial eye, the conditions I must take into account in my preparation are totally different from those that Nature faced billions of years ago when she set out to make organisms. "Nature," of course, is only symbolic shorthand here, because there was no constructor then to gather appropriate molecules and keep fitting and joining them until the first bacteria popped up. Nothing like that ever happened; there was only a primordial, warm ocean with dissolved organic and inorganic salts in it, and nothing else. As we know today, biological evolution in its proper sense was preceded by a long evolution of organic molecules or, more precisely, an evolution of chemical reactions through their mutual competition and "natural selection." In the course of many simultaneous reactions, large molecules composed of tangled threads of atoms, called polymers, came into being, and at a certain stage of this process, they separated from the ocean in the form of tiny colloidal droplets. This phenomenon resulted from the operation of the fundamental laws of physical chemistry, and we can readily reproduce it in the laboratory. The droplets were not yet cells, but cells arose from them in the course of the subsequent "chemical evolution," perhaps a billion or several hundred million years later. Please note that the colloidal character of the protoplasmic droplets, the future building blocks of multicellular organisms, was determined at a very early stage of evolution, since the "natural selection" of chemical reactions could not take place anywhere else except in these droplets which constituted a concentrated, reactive phase of specific groups of compounds. This was the beginning. Afterward, the plasma adapted to changing conditions, but various features of the structure and functions of our body indicate that life originated in the ocean whose waters were as salty as our blood. Nature built organisms where it was possible—in this case, in the water—because at temperatures prevailing on Earth, *corpora non agunt nisi solute*, compounds do not react unless they are in solution.[1]

Further, Nature built from what was at hand: in the ocean, some compounds were abundant, some present only in traces, and some totally absent, and the composition of our body reflects that. But today's engineer who wants to build a brain prosthetic has at his disposal much more than aqueous suspensions of sticky colloids at relatively low temperatures: he can use various kinds of machinery, rare substances, high temperatures and pressures, and so on. My point is that the structures of our body and brain reflect not only the biological purposes that they have been serving up to now but also the monumental, long, and complicated path of the entire biological evolution. That is why we carry traces of both the early prebiotic phase and the later formation of living organisms in biological evolution through natural selection, changes in the environment, and inter- and intraspecies competition. Furthermore, evolution was in fact not a ceaseless upward movement; there were stumbles and defeats. It improved and perfected structures and processes but also regressed, degenerated, and eliminated forms and species. The road from unicellular organisms to human beings was full of zigzags, wrong turns, and blind alleys, and our body carries traces and consequences of those "tactical maneuvers" of the evolutionary process even today. Of course, the engineer of the brain prosthetic or electronic brain does not need to concern himself with those zigzags, those traces of the long-gone evolutionary stages or the vestiges of adaptation to conditions in which the prehistoric ancestors of humans lived. Because I like digressions, I add that a limited lifespan is not a consequence of some "construction error" on the part of evolution; it is the consequence of an engineering necessity, for evolution is driven by the variability of forms and their succession. Where there is no death of some forms, room is not made for others to follow, and there can be no evolution. Individual death is the price we pay for the possibility of the continued development of our species. But to return to the topic: biological evolution was the only possible way to reach and cross the threshold of the minimum complexity mentioned earlier. The path from aqueous solutions to colloidal droplets to cells to multicellular organisms to humans was the only path allowed by the conditions on our planet. Once the threshold of the minimum complexity was reached and organisms became protected against deleterious effects of degenerative tendencies, the proper organismal evolution could begin. But of course, none of this interests the engineer of prosthetics, artificial hearts, kidneys, eyes, or brains. Nature faced many difficult conditions, but the constructor today encounters just a few. And this is the reason why the engineer of the future will be able to construct a conscious machine from glass or metal and not bother with sticky proteins in solution.

HYLAS   You really think that life and consciousness are not inseparable and that there may exist a structure made from nonliving elements that can house consciousness?

PHILONOUS   Definitely. I also think that the fundamental rules of the brain's functioning must be the same in all corners of the material universe, though other beings who have brains may differ from us as a star differs from a starfish. Their mentation will still share the same principles of induction, deduction, and Ockham's razor (a sparingness of hypotheses).

HYLAS   I have a few more questions about consciousness.

PHILONOUS   Go ahead, my friend.

HYLAS   I accept your thesis that consciousness is a group term that applies to an entire class of phenomena. But where, in your opinion, is my consciousness located? In my head?

PHILONOUS   Where else could it be?

HYLAS   Can you then point it out to me with a finger?

PHILONOUS   My digestion takes place in my abdomen. Can you point it out to me with a finger?

HYLAS   I could show you, and the world, your digestion processes by surgically opening your abdomen under local anesthesia. But if you open my skull under local narcosis and show me (in a mirror) my brain, neither you nor I will see my consciousness. We will not even see any of the processes that together constitute this group term. Because it is impossible to see my perceiving or thinking of clouds or my toothache. So I submit to you that consciousness is not located in physical, objective space at all. If we accepted that consciousness had a location we would arrive at some very funny notions. For example, when I bow before a meal, my hunger also goes down, or when, suffering from unrequited love, I bang my head against the wall my love also periodically hits the wall?

PHILONOUS   Please, Hylas, tell me where in physical space "the repulsion of bodies with equal charges" is located?

HYLAS   I see. "The repulsion as such" I cannot show you, because it is an abstraction. But I can show you the direct act of mutual repulsion of two equally charged bodies.

PHILONOUS   You think so? You will show me only how the distance between the two bodies increases. One can see the movement but not the repulsion. Repulsion is a generalization, an abstract term, like love. If we elevate the laboratory bench on which you perform your experiment with the equally charged bodies, will you say that the repulsion was elevated too? Or, for another example, can you see an electron?

HYLAS   Of course, in a Wilson cloud chamber or on a photographic film.

PHILONOUS   Not at all. In the cloud chamber you only see a streak of vapor condensed on the ions that something has formed, and on the basis of atomic theory you conclude that that something was an electron. And on the photographic film you see only a few blackened grains of an emulsion. You do not see an electron directly; you always only deduce its presence on the basis of certain traces and physical theories. Similarly, when neurophysiology progresses sufficiently, I will soon be able to show you certain electrochemical processes in your brain and on the basis of which I deduce that you are seeing, hearing, or thinking (e.g., thinking of a special person "with tenderness and devotion," which is an expression of love). A distinct group of processes take place in your brain when and only when you are sad, and so sadness exists in the same way that consciousness does. Both are abstractions encompassing a series of phenomena that are interconnected and therefore can be assigned to the same set.

HYLAS   I am not convinced. A person can feel sadness or a toothache directly but cannot experience or feel electrostatic repulsion or an electron.

PHILONOUS   Consider what is happening when you feel hungry. Why are you hungry? Because your empty stomach sends signals to your brain, no?

HYLAS And so? Those nerve impulses can be observed by any observer with a galvanometer, but it will not make the observer feeling my hunger. That is my private perception, as opposed to the publicly observable nerve impulses from the stomach to the brain. Don't try to erase that difference, please.

PHILONOUS I am erasing nothing. Look into your belly, and you will see, thanks to the information traveling through the nerves from your eye to your brain, that your stomach is performing certain movements called "hunger pangs." And you feel hunger because of the information traveling, through other nerves, from your stomach to your brain. The only difference is that the information about hunger is addressed only to *your* brain, because your stomach is connected to no other. In contrast, anyone can see your stomach in your opened belly. Other people, of course, could only see, not feel, the pangs. But if we connected the nerves from your stomach to my brain, it would be me who feels the hunger even though the empty stomach is yours.

HYLAS You posit an unnatural procedure.

PHILONOUS Nonsense, Hylas! Connecting your nerve with mine is an "unnatural procedure"? Then using an electron microscope to study atomic structures is an unnatural procedure too. In both cases we are conducting an experiment to confirm our assumptions and gain new knowledge about the world (which includes both objects around us and inside us). If you forbid scientists to act "unnaturally," we will have to limit ourselves to satisfying our hunger, thirst, and sex drive. Nothing beyond that, because only that is "natural." You cannot be serious with this objection.

HYLAS What an outburst! Very well, I withdraw my objection regarding the "unnatural." Continue.

PHILONOUS We have thus established that the distinction between "private" and "public" facts boils down to the relation between a given person and given information. Information about what is happening inside the person's body is directly accessible (through the nerve connections) only to that person. Information from outside the person is directly accessible to everyone present. That is all the mystery.

HYLAS Let me repeat. You are saying that the difference between a subjective perception ("I am hungry") and an objective perception ("I see a photograph") reduces to the relation between the information and its addressee. Information about internal processes is fed through the nerves exclusively to the brain of that organism, while information from the surroundings is accessible to everyone.

PHILONOUS   Yes. From this it also follows that any particular information may reach your brain in two ways if it originates in your body. You can *either* observe your stomach with your eyes (after we have opened your belly) *or* "feel" it, that is, "feel its emptiness" through the direct nerve connection. Needless to say, this distinction results from evolutionary adaptation, for it would be useless, or even harmful, if we felt someone else's hunger or toothache.

HYLAS   But what is the source of the information that I am sad, that I am experiencing sadness?

PHILONOUS   That information is a message to your brain about its own state through the system's "internal feedback." I think that now we have dispelled all doubts that anyone can have on this subject and are ready to tackle the main problem: the functional analysis of a system belonging to the set of "feedback networks" mentioned earlier.

HYLAS   Cybernetics studies such systems?

PHILONOUS   It does. But the topic is vast and challenging, and tackling it requires considerable intellectual effort. So let us postpone it to our next meeting. In the meantime, please, think about the issues that we covered today, above all—evolution viewed from the constructor's standpoint.

V

HYLAS My friend, I spent all night thinking about what we have considered so far and would like to hear your answers to a few questions before you continue to show me how the discoveries of cybernetics have relevance to the philosopher. If I understand you correctly, you make the distinction between "subjective" and "objective" according to the way information is "inputted" into the nervous system. Because my stomach is "plugged" into my brain by my nerves, I can experience hunger directly. Because objects outside me are not "plugged" into my brain, I can perceive them but not "experience" them directly. But things get complicated when my own brain is the perceived object. I can experience it both "from inside" (directly) and observe it in a mirror after a hole is drilled in my skull (under local anesthesia). Only I have the first kind of access to my brain; everyone else has only the second kind. How to reconcile this duality?

PHILONOUS There is no duality because your brain "experienced from inside" as you put it, does not exist. Thinking and feeling *take place* in your brain, but *they are not your brain*. Your brain "is," that is, it "exists" only in the way that outsiders observe it.

HYLAS Could you be more precise and define consciousness in objective terms?

PHILONOUS But of course. Consciousness is a characteristic of a system that one can recognize if and only if one is that system. This statement contains both necessary and sufficient conditions for calling something consciousness in an entirely objective manner. If you listen to another person on the phone, it is only you who hear the person, unless we plug another listener into the line. In the same way, if you could plug yourself into another person's brain, you would directly participate in that person's circulation of information, that is, his mental life. We will talk later about how such an experiment might be done. What is your next question?

HYLAS Despite all your explaining, I still don't know what consciousness really is.

PHILONOUS To tell someone what something is means to equate that "something" to "something else" and build a model of that "something else," mathematical, mechanical, or another, and that's all. We have no other way of knowing or understanding in this vale of ours. Therefore, it is not clear to me what you mean by "real" understanding.

HYLAS I suppose an extraordinarily complex electronic brain could manifest behavior indistinguishable from that of a human being: it could perceive and study not only its environment but also itself, it could think, it could express its thoughts, it could reason—but we still would not know if all this activity was accompanied by consciousness. As you say, the only way to find out would be to become that electronic brain.

PHILONOUS You are in bondage of superstitions and obsolete prejudices to a degree that drives me to despair, Hylas. You keep saying that you don't know what consciousness is, and then suddenly you claim expert knowledge of the topic, which has undoubtedly sprung from some kind of heavenly revelation.

HYLAS What are you talking about?

PHILONOUS What you said indicates that you still believe that consciousness is not a generalization of processes like thinking, perceiving, and feeling but a kind of absolute that oversees and nurtures all those processes but cannot be reduced to them, and therefore is a superphenomenon or *epiphenomenon* that hovers over mental activities like the Spirit over the waters. You have turned out to be an epiphenomenalist, Hylas. In vain have I repeated that consciousness *is* the seeing, *is* the hearing, *is* the feeling, perceiving, remembering, learning—and nothing more. You yourself reached that conclusion when you agreed that consciousness consists of mental processes as an army consists of soldiers, but now a mystical revelation returns with a fresh blush of metaphysics on its cheeks, in full strength and health.

HYLAS   You are right. I did not express myself well. But . . . if you equate consciousness with an organism's (or electronic brain's) reaction to stimuli, you are erasing any difference that may (and in my opinion must) exist between a lifeless thinking machine and a living human being. *My* perceiving, *my* thoughts—that is *my* consciousness, fine. But does it necessarily mean that an electronic brain's perceptions and thoughts are *its* consciousness? In the jump from a human to a machine, my mental processes somehow lose their inner quality.

PHILONOUS   The answer to this quandary lies only on the path of experimentation and empiricism. You acknowledge the inner quality of your own mental processes and do not deny other people the same quality, because they are constructed like you and from the same material. I fully understand your resistance in the case of an electronic brain, but perhaps that can be overcome if I show you how you (or any other person) can become an electronic brain. Then we will see whether or not the inner quality of mental processes disappears during that transition.

HYLAS   But this is absurd, impossible!

PHILONOUS   Form that opinion only after we have collected enough data and have sufficiently penetrated cybernetic science to propose an appropriate experiment.

HYLAS   Very well. Are you going to speak about the set of systems called networks?

PHILONOUS   Exactly. This set contains systems whose degree of complexity is equal to or greater than w, which signifies the minimum complexity that a system must have to belong to the set.

HYLAS   And all these systems possess consciousness?

PHILONOUS   If we define consciousness as a system's feature that we can directly recognize only when we are the system, then we would have to ascribe consciousness to the brains of reptiles, birds, fish, and even to the "abdominal brains" that are the ganglia of insects. However, such broadening of the meaning of consciousness is inappropriate.

HYLAS   So your definition fails?

PHILONOUS   No. We just amend it: consciousness is a system's feature that we can recognize only when we are the system, and the system's complexity approaches that of the human brain. This judiciously narrows the scope of the term. As for the brains of other animals and the networks that are not brains of living organisms, we can only assume that an equivalent of human consciousness appears to various degrees: the higher the network's organization, the "higher" or "clearer" its consciousness. The fuzziness of this formulation stems from the fact that as yet we cannot measure consciousness by physical means. Such measurement is possible in principle or in theory, but its practical realization is still far off.

HYLAS   How do you envision it?

PHILONOUS   It will undoubtedly measure the amount of information, that is, the "opposite of entropy," but it must also take into account all the *transformations* that this information may undergo in the given network, as well as the network's effects on its environment and *vice versa*. If the transformability of information is a function of a network's complexity, and there are many indications that this is true, then we will be able to derive mathematical equations for the relation between the complexity of a system and the degree of consciousness that it can possess. In our set there is a hierarchy of networks, from the simplest, which barely reach w and manifest very "weak" consciousness, to the most complex, with a complexity of w^n, whose consciousness is the "highest" and "clearest." With such mathematical tools we will not have to resort to such imprecise terms for consciousness as "clear," "dim," "low," or "high."

HYLAS   Wait. Your statement has one curious implication. You are saying that in this set of networks, the simplest are at the bottom of its hierarchy, while the most complex are at the top. But complexity can be arbitrarily increased, therefore consciousness has no limit. An infinitely complex network would have "infinite" consciousness. The mathematical expression of this thesis would be equivalent, I believe, to a formula for God, a being with an "infinite" consciousness.

PHILONOUS   What you say is amusing, but it is not like that. In addition to the threshold of the minimum complexity there must exist a limit of the maximum complexity.

HYLAS   What determines this limit?

PHILONOUS When consciousness increases beyond a certain point, most likely it will cause regress and degeneration.

HYLAS How so?

PHILONOUS Exceeding the optimum complexity, a network begins to functionally deteriorate: its separate parts begin to liberate themselves from the central integrating forces; autonomization tendencies lead to internal conflicts among the processes; and finally, the excessively complex network breaks up into quasi-independent units engaged in mutual battles, to the detriment of the whole.

HYLAS This sounds like a fantasy.

PHILONOUS It is not. Of course, we do not know if the human brain approaches this limiting value, that is, if it has already passed the level of optimum complexity, but under certain conditions it exhibits clear tendencies toward autonomization of its parts. In the functional, not material, sense.

HYLAS What are the symptoms of that, and why do you say functional and not material?

PHILONOUS The disintegration of processes and the related degrading of a network's functional integrity (that is, its "personality") do not require a physical division in the system. A personality split in a person may reach an advanced stage without any morphologically or anatomically detectable changes. There are probably many factors involved in the determination of the optimum complexity, such as the speed of impulse transmission or the number of "degrees of freedom," which, in turn, is a function of the number of possible transmission pathways.

With the functional autonomization that complex networks exhibit, we enter the vast realm of mental phenomena related to the so-called subconscious. It is a field of psychology that, more than any other, is plagued by murky terminology and a flood of unverifiable hypotheses whose originators typically are methodically incompetent or untrained followers of Freud. For this reason, the cybernetic analysis of the subconscious and research in this field based on the directives of information theory are especially needed.

Attacking the problem of the subconscious with the tools of cybernetics, I will address only human mental development, not comparing its neuronal network dynamics with that of other network types (e.g., electronic), saving what I would call the engineer's approach for later. Psychoanalysts say our mental life contains two parts and is a resultant of two kinds of processes: conscious (personified by the ego, that is, the conscious "I") and subconscious (the so-called id, a collection of mental phenomena that a person normally cannot access).

Conscious processes are distinctly and obviously adaptive and *purposeful*, that is, they are manifestations of the causatively expressible and biologically rational adaptation of the human organism to its environment; they arise on the path of *learning* all the network activities necessary for survival. The processes that are ineffective or do not lead to the achievement of this goal are, thanks to negative feedback, inhibited and eliminated from the organism's behavior.

The subconscious processes lack such an objective- or goal-oriented, rationally causative character. They appear oriented toward principially unattainable goals that are, moreover, irrational and lacking any biological purpose. They are typically expressed by the perseveration (a circular repetition) of activities characterized as obsessions, phobias, neuroses, and so on, which do appear in the behavior of so-called normal people but are manifested with greater force and clarity in the behavior of mentally disturbed people.

What is the origin and mechanism of these processes? A newborn has at its disposition a network in which the majority of processes are scattered, random, and aimless. The result is disordered muscle movements, chaotic variability of responses, and the inability to act in an integral, purposeful way. As the baby gathers experience, it starts *eliminating* aimless activities (i.e., those that do not lead to the achievement of any goal). The initially chaotic system whose behavior reflected "statistical randomness" of its network processes begins to organize itself in distinct functional sets corresponding to particular tasks. The child learns how to look, that is, to turn the eyes in a desired direction, to walk, to speak, and so on. The "statistical randomness" of the "newborn network" must be understood *cum grano salis* lest we fall into a kind of "physical absolutism" and treat the functionally unorganized network like a random collection of atoms, since clearly the network has some "centers of functional crystallization" from its inception and its activity is not as disordered as the motion of a swirl of Brownian particles in a drop of water.

The progress from random activities to those that exhibit a group character, from thoughts that are foggy and equivocal to concepts that can be precisely expressed in words, and from newborn behavior to that of an adult takes place through learning and the elimination of purposeless processes, through specialization, organization, and dynamic structuralization, in which the selection criterion is the adaptive success of a given mental activity or its effects when transposed into action.

Each new skill requires the full concentration of consciousness, that is, all those higher-level network processes that the psychologist calls the "personality" or "ego." Once a skill has been mastered, it moves from the conscious to the subconscious region as an automatism. No longer must the entire network build a dynamic model of the activity to achieve the adaptive effect, no longer is it necessary to focus the attention at each step of the new action—because a special functional subsystem has been formed that is always ready to initiate it whenever it is called for. This manner of automatization and transfer to the subconscious applies to all functions without exception, from the aiming of the eyeballs (not easy for a newborn!) to the most complex activities, whether kinetic (acrobatics and juggling) or mental (abstract mathematical reasoning may be beyond a layman but is executed *automatically* by a professional mathematician).

All automatisms of this kind share one characteristic: they can be retrieved into consciousness at will. For example, the automatism of breathing, riding a bicycle, or acrobatics can be moved to the center of the subject's attention, which then allows for an introspective study of parts of the activity.

The automatisms of the subconscious, however, differ from those of the unconscious in that they are not freely accessible, and any attempts to trace their sources meet with considerable difficulties. The reason is that a special *dynamic barrier separates them from consciousness*.

Both conscious and subconscious phenomena are based on a symbolizing (symbol-generating) function of the network, but the application of this function is fundamentally different. Conscious symbol formation uses symbols as abbreviations or "call codes" of large sets of impulses to create "situational models" of the external world or of particular states of the network itself. The creation of such models and operating on them (e.g., the transformation of a thought into words or into a mathematical formula) are necessary for the human organism's adaptive functions. Such symbols mainly address the surrounding world and serve the purpose of communicating with other people; they also help create a "model" of the external world within the network. The biologically regulated, rational, causatively conditioned, and adaptively indispensable function of this group of network processes—note that all this is conscious—is self-evident and well understood.

A substantially similar symbol-generating ability exists in the subconscious, which encompasses mental processes that do not constitute automatisms in the sense described above (because of the "dynamic barrier," there is no access to them), and their adaptive value is questionable. Granted, their inaccessibility is not absolute, since we know of them. They manifest themselves in dreams, hypnosis, and many disease states of the mind. They also can be uncovered by tests, particularly by the method of "free association."

The symbolic functions of the subconscious are "forced" onto consciousness when the network is impaired; they take the form of obsessions, tics, and phobias and display great persistence, along with perseveration tendencies and a resistance to arguments of experience or rational persuasion, whether they come from the subject himself or from other people. These symbolic functions have all the characteristics of "normal," purposeful network activities, namely: (1) an initial motivation, (2) an ensemble of activities, and (3) a goal. But the elements, when taken together, form a completely irrational whole that does not serve the subject or anyone else and is the very opposite of adaptive—it torments the person who is driven to such actions by the inner imperative.

We have here a reversal of the hierarchy of the mental phenomena that we consider rational. We know that consciousness is to some extent conditioned by many unconscious mental processes—for example, an articulation of a thought is enabled by a considerable number of mental automatisms, such as the internal memory feedback connections that supply the necessary words, maintain the articulation's directional gradient (we always speak "about something," but at the same time our mind is moving "from somewhere" to "somewhere else"), eliminate from the field of perception and thinking all stimuli that might interfere with this articulation, and so on.

The phenomena of the unconscious (which can always be retrieved into the conscious) are, so to say, the foundation of the edifice of consciousness. One expects the relation between the two domains will be one-way, that consciousness will retain full control over the unconscious automatisms that support its functioning and that those automatisms will be retrievable as needed. The reverse, that is, the control of consciousness by the unconscious automatisms, would at first glance appear to be impossible. But the relation between them is much more complicated.

First, let us recall the random behavior of the "newborn network." The network is an "apparatus for generating hypotheses" about the surrounding world and about the organism's relation to that world. It generates—by "trial and error" and elimination, as we know—models of the environment and dynamic models of purposeful behavior. The main point is that not all "incorrect models" are eliminated completely; some leave traces that can last a long time. The factor that enters the game at this point is the organism's (the network's) emotional engagement with situations that, because they are incorrectly modeled by the network, should be removed from the conscious. Yet they remain in the network as subconscious processes, hidden behind the "dynamic barrier," from where they can affect in a harmful way the course of other, conscious processes for years.

A phenomenon that I am going to introduce next may shed some light on these issues. I mean *memory* and the mechanisms of remembering and, more importantly, of recalling at will. The human neural network is not as efficient, reliable, and subordinate to consciousness as some think. It has been shown experimentally that human memory contains about a *million* times more stored information than the recollection mechanism can access. Under hypnosis a person can remember things that he could not while awake. American psychologists have shown that hypnotized bricklayers can describe in detail a few bricks, out of the tens of thousands, that they laid at a building site eight to ten years ago. A man described a particular brick in the wall of a building that he put up—the sixth from the end in the eighth row on the third floor—as having a reddish spot at its edge and a small chip at the left corner, and this was confirmed! These amazing results show that the memory of an average person contains about 10^{15} elements, of which no more than a millionth can be consciously accessed when the person is awake.

Today we can only guess what the connection is between these facts and the phenomena of the subconscious. It is possible that the subconscious represents a set of phenomena that inevitably appear in a network that has crossed the complexity threshold. It is a side-effect that evolution did not intend, so to speak, yet it is indispensable for the functioning of sufficiently complex systems that possess consciousness.

As we can see from the experimental data, we have very limited control over the stores of our memory. In those stores, various processes of which we know little—for example, the spontaneous formation of connections and the transformations of specific engrams (memory traces)—take place. This gives rise to certain sets that, though hidden behind the "dynamic barrier" of the subconscious, affect the totality of consciousness in complex ways. Today we can sketch hypothetical maps of neural connections that seem to correspond to the movement of engrams from the retrievable region of recollections to the sets that are the substrate of the subconscious. These are the first steps of cybernetic analysis in this difficult terrain. The method of free association allows us, so to speak, to "send a probe" into the subconscious and obtain "samples" of its raw material, whereas the analysis of the conscious content is limited to what has been already filtered through the dynamic barrier and undergone organization, for which reason it tells us little, if anything at all, about the phenomena occurring outside consciousness.

To conclude, consciousness would not be possible without the unconscious, automatized processes. And without the symbol-generating function there would be no subconscious that manifests itself in our heads in dreams, in hypnosis, and in neurasthenic but also normal states—in the form of mistaken actions, the forgetting of a name or word, and so on.

HYLAS So we don't understand the symbol-generating function of the subconscious or the purpose it serves but suspect that it may be a side-effect of the functioning of a very complex network-type structure?

PHILONOUS Correct. With this, let us finish our digression on the topic of psychoanalysis. We used it to underscore the fact that the relative freedom or autonomy of certain mental phenomena is the fundamental functional characteristic of a large neural network. Nevertheless, cybernetics cannot tackle this problem with sufficient rigor, because as a science it is still in an embryonic stage, and a general mathematical theory of automata that are arbitrarily complex networks does not yet exist.

HYLAS What theory do you mean?

PHILONOUS The most complex networks that we construct today are computers. They contain 3,000 to 4,000 elements (crystalline or vacuum relays). The largest networks that we can assemble using the current technology and knowledge are limited to about 10,000 functional elements. The number 10,000, or 10^4, is thus their complexity coefficient. In contrast, the network of the human central nervous system contains 10^{10} elements (neurons), which makes it a million times more complex than the largest "artificial" network. Why are we unable to create automata larger than 10^4 elements? The first difficulty is technical, the size of the vacuum tubes and their relatively high power consumption: a vacuum-tube electronic brain 100,000 times larger than the existing ones would require the entire Niagara Falls to cool it.[1]

But there is progress on this front. Using transistors instead of vacuum tubes, we could cut the size and power consumption of an electronic brain by ninety percent. The second difficulty is our ignorance of a general theory of automata. The first airplanes were built empirically, by the method of trial and error; but the further development of aircraft would not have been possible without the theories of flight, aerodynamics, material strength, induced vibration (flutter), and so on. We still construct networks essentially by trial and error today, as there are no generalizing theories of how they function. The theory of digital automata, such as computers, must become a part of formal logic. The first problem is that formal logic does not tell how many elementary operations are needed to solve a task; it only says if the problem is solvable or not. For formal logic it is irrelevant if a solution would require so many operations that performing them would take ten billion or a quadrillion years. But in automaton construction one must know the number of operations needed to complete a task. The second problem is that a network can make a mistake in any of its elementary operations. When the number of operations becomes enormous, the probability of error increases. To minimize the effects of errors, an organism employs corrections that we cannot: the tissues of an organism are self-repairing, but technological systems do not have this ability. A future theory of automata therefore must consider the length of the reasoning chain, which in turn must take into account the factor of *time*, and it also must consider and predict the rate of production of reasoning errors. In order to do that, the theory must, first, join logic with thermodynamics (which includes the time factor in the processes of entropy increase, for the opposite of entropy is information), and second, include the data from biophysics.

HYLAS I do believe that such a theory will enable us to determine the optimum complexity, beyond which the wholesome functioning of a network begins to break down. But I still cannot see how a theory that is physical in nature and therefore expressed in the language of mathematics can measure the degree of consciousness in a given network.

PHILONOUS It will determine the efficiency, scale, and rates of all those processes that constitute consciousness, but nothing more. I will give you an example of what such a theory might do. On the basis of their analysis of certain neural networks, McCulloch and Pitts concluded that if one briefly touches a person's skin with a cold object, the person will sense heat.[2] Their prediction was confirmed by an experiment. We already can find mathematical equivalents of the functioning of simple networks; for very complex ones, this is still impossible.

HYLAS Yet none of this addresses the issue of the "inner quality" of mental processes.

PHILONOUS   Why do you think so? Surely you know that human consciousness is variable and not always equally "clear." Certain chemical compounds may "sharpen" it, while others blunt it. The consciousness of a person awake is different when he falls asleep, and still different when illness or fatigue affect it, there are "states of obtundation,"[3] and so on. The theory will predict all these possibilities. But perhaps this is not your point? Perhaps, by "inner quality," you mean what the brain of a fish or the ganglia of an insect exhibit, that is, metaphorically speaking, "what it feels like to be a carp or an ant." A theory of automata will, of course, be unable to put us in the shoes of a carp or an ant.

HYLAS   Here's another objection, which concerns all of cybernetics, but especially its philosophical fundament. "Feedback networks," which are the subject of this science, are mechanisms. So cybernetics, reducing neural and even mental phenomena to mechanical phenomena, is merely a new reincarnation of the nineteenth-century mechanistic materialism, which held that everything, including all the processes of life, could be expressed in the language of mechanics. But haven't advances in biology and physics torn down the edifice of that naive idea?

PHILONOUS   You are saying that cybernetics is a continuation of the old mechanical philosophy. But consider this. Philosophy is always a reflection, an abstraction, of the practical activity of human beings. At early stages of their existence, people, having already coalesced into groups, the nuclei of society, and having acquired language, attempted to affect the surrounding world and explain it in the same way that they used with respect to other people. Their personification of natural phenomena, heavenly bodies, stars, and so on, was the first general model of those phenomena. It was both anthropomorphic and animalistic. At a much later stage, in the seventeenth and eighteenth centuries, a new model of phenomena started to form. Its basis was a mechanism—a human artifact in the form of a clockwork, a machinery—together with its theoretical underpinning in Newton's celestial mechanics. Physics started to treat matter as a collection of tiny elastic particles subject to laws of mechanics. The laws of mechanics succeeded in solving the mystery of the heart and the circulation of blood. Mechanics helped to create the first steam engine. The underlying concept of a mechanism deduced from all these fields had the following characteristic properties: the whole could be reduced to the sum of its parts, every process could run equally forward and backward, and the mechanism was ahistorical, that is, unaffected by its history. You can take a mechanism apart and put it together again without changing its function. You can reverse its operation and it will return to its starting point. When you know the position of all its particles and the forces acting on it, you can predict its arbitrarily distant future. The problem is that these statements are true only when applied to systems like clockwork or a steam engine; they are not when applied to biological or quantum phenomena. Experience has refuted all the postulates of mechanical philosophy: an organism is more than the sum of its parts; the processes occurring within its boundaries are irreversible; its history does affect its future, and its future states cannot be completely predicted by knowing its past states. Mechanism therefore had limited value as a model of the phenomena that take place in nature, especially those in living systems (or more generally, in organisms, whether living or nonliving, as defined by our threshold of the minimum complexity). This, of course, does not mean that this concept did not play a positive role in scientific progress in its time. We only must take care not to fall into the trap of mechanical philosophy. Cybernetics rejects that model and offers a new one: a system that is not reducible to its parts but instead represents a distinct, unified whole; a system that is shaped by its past developmental history; a system that actively coexists with its environment; a system whose future behavior cannot be precisely predicted from its structure. If you insist on calling this new system a mechanism, then you must apply that term to living beings as well.

HYLAS So you are saying that a so-called complex network is not a mechanism?

PHILONOUS It is a matter of word choice. We shall not go into details of network technology or network evolution, that is, the causes of their emergence. We only say that regarding their material aspects, networks can be made of tissues, electrical conductors, mechanical blocks, or coupled chemical reactions. Thus we can have networks that are called neuronal, electronic, mechanical, chemical, or even combined, whose individual parts are made of different material. As for their evolution, let us note that the networks of living organisms emerged in biological evolution, but all other that we know are results of human engineering.

HYLAS In your opinion, then, an electronic network could form consciousness? But how can we prove this if consciousness is an inner quality of a system that we can perceive only when we are that system?

PHILONOUS The question of whether or not a network possesses the "inner quality" of conscious perceiving shall be left open. This is irrelevant to cybernetics. The purpose of cybernetics is to construct networks that manifest phenomena such as "memory," "learning," pursuing "goals," "recognition," satisfying "needs" (which can be either "real" or "substitute"), forming "habits," having "obsessions," and establishing "values" such as "free will," "personality," "freedom of choice," "initiative," "creativity," "character," "temperament," but also "neurosis" and "addiction," and others.

HYLAS These properties would emerge in the devices that you call networks?

PHILONOUS Obviously not in all of them. We will consider several different kinds of networks.

HYLAS Go on. I am all ears.

PHILONOUS The fundamental feature of all living beings is their teleological or goal-oriented character. Every life function of every organism is subordinate to a goal, which is the continuation of life, of both the individual and the species. This statement is obvious and banal, because by living, human beings, lions, hippopotamuses, or flies serve no purpose beyond themselves. It is a different story with machines, the products of human beings. The goal of a machine's existence relates not to the machine but to the realm of human activities. Thus a microscope is an amplified human eye and a steam engine or atomic reactor an amplified human arm. A machine also enables human beings to perform actions that they could not perform without it, such as flight, by an airplane. In each case, a temporary or permanent human intervention to correct and regulate the machine's action is necessary. All machines are connected to the human nervous system, which steers them (the pilot in the airplane, the engineer in the train, the physicist at the atomic reactor). Human beings can also make use of the life functions and properties of a variety of other organisms (both animals and plants) for their purposes. It is no different when they build feedback networks. The networks created by constructors have no goals of their own that would relate exclusively to themselves: an electronic brain helps people do calculations; an autopilot helps steer an airplane; and so on. Every such network, being a creation of its constructors, imitates the activity of the human nervous system in a specific, very narrow field, not the activity of the system as a whole but only one of its parts, separated from the rest. For this reason, analogies drawn between a computer or other network of that kind and the human brain will not take us very far. But that may not apply to the artificial networks of the future. Here I intend to preempt the accusation of paying too little attention to the networks that exist and operate today. The reason is that we are interested not in the practical uses of specialized networks but in the properties that we may find in networks in the future, properties that today we detect only in the activities and operations of the human brain.

After this clarification, we can finally get to the point. Every network must be in contact with its environment, and to this purpose it is equipped with devices that allow information input and output (which also can include an effector that acts upon the environment, as we see below). The information leaving a network may or may not be translated into physical action. The electronic brain designed for calculation receives information from its environment (the operational instructions and the assignment), and its output releases information "transformed," a result of mathematical operations. But a device for shooting down airplanes, consisting of a radar (input), a network (the steering unit), and cannons (output), also possesses an effector (the cannons).

Since the computer is like an isolated part of the brain into which we must feed information and then take it out, transformed, and since the computer performs no physical activity directed at the environment, let us focus on the operational principles of the antiaircraft apparatus. It can "perceive" the airplane through the radar, "recognize" it as an airplane (as opposed to a leaf in the wind), then, on the basis of previous experience stored in its "memory," predict the probable position of the airplane several seconds later, and finally, aim the cannons at that position and shoot the airplane down. The network can make a mistake—in "perception" (taking another flying object, such as a kite, for an airplane), and it also may "miscalculate" when the airplane performs an improbable maneuver to change the direction of its flight abruptly. If the first round does not hit the airplane, the network recalculates and shoots again, repeating this cycle until the goal is achieved (the airplane is hit). When two airplanes appear at the same time, a "conflict" ensues, and the network must "decide" which airplane to shoot at first. If it works as expected, the network will "decide" and start firing in the described manner. For each subsequent shot, the error (the deviation between the line of fire and the airplane's position) is corrected by feedback and is smaller and smaller until the target is hit. If the network does not have a device for making decisions in the event of a conflict, it will not be able to "decide," will "hesitate," and there will be a series of alternating processes (making a decision, changing its mind, changing its mind again, and so on). So I ask: What is the fundamental difference between such an apparatus and a living organism?

HYLAS   The organism is living; the apparatus is not.

PHILONOUS   This difference is indisputable but for us unimportant.

HYLAS   The organism's activity aims at continuing its own existence, while the activity of the apparatus that you describe does not.

PHILONOUS   Precisely. Feedback enables the network to approach the target by continuously correcting its behavior on the path toward the goal, but it does not set that goal. The goal is given: for an organism, it is set by evolution, and for our network, by its constructor. One can say that in general the "primary cause" for directed action is provided to a network by its inner imbalance. The imbalance can be biochemical, electrical, or mechanical. The goal for the network to achieve is to reach equilibrium. Giving an electronic brain a task, starving an animal, or flying an airplane through a radar field disturbs the inner equilibrium of the network (computer, animal, antiaircraft automaton), and this disturbance initiates certain patterns of activity. When the electronic brain has solved the problem, the animal has obtained food, and our automaton has shot down the airplane, they all reach equilibrium, which will last until a new stimulus, external or internal, disturbs it again.

HYLAS   You are talking about objective goals, not about those that only subjectively exist in consciousness, right?

PHILONOUS   Yes. I am talking only about teleological processes that are objective. Our antiaircraft network, the computer, and the animal are not "aware," in the human meaning of the word, of the goal they are pursuing. But there always exists an event or series of events that eliminate or minimize the imbalance and restore equilibrium. Such an event is usually external to the network (the animal finds food or a sexual partner, the antiaircraft system shoots down the airplane) but also can originate in the network itself, as when the network recombines its own elements to lessen its imbalance.

HYLAS   Can you give an example of such a pursuit of a goal that only exists within the network?

PHILONOUS   Such regrouping of a network's internal elements occurs when a poet writes a poem. The goal is "internal," as is the action—putting word symbols into a new structure which lessens the network's imbalance.

Sometimes achieving the goal is beyond a network's ability. It can happen that a network, failing, will start to pursue a substitute goal. This phenomenon shows in an electronic brain as a short circuit and in a human as drug addiction; in other words, a surrogate activity takes over. These are pathological states. In the extreme, they lead to the self-destruction (a suicide) of the network that cannot solve the tasks it faces.

HYLAS   Since you are not using moral values in your analysis, why do you call the poet's poem an achievement of the goal and drug abuse a pathological phenomenon?

PHILONOUS   A very good question. The drug's effect will lead to an internal recombination of the network that accomplishes a decrease in imbalance. But the poem represents an *increase in information*, the drug use does not. In general, one can call pathological any internal transformation of a network that lessens its ability to achieve real goals.

Let us now consider the second fundamental feature of a network: the ability to learn. This ability presupposes the possession of memory. Learning is a modification of behavior based on data from previous experience—thanks to feedback. In the language of cybernetics, it is an internal regrouping of a network's elements for the purpose of pursuing a goal more effectively. The simplest mechanism of learning is the formation of temporal associations (conditioned reflexes). Each network has "rules of internal traffic" or "a system of preferences." At a street intersection, certain vehicles have the right of way, and the total traffic is conditioned by the configuration of the intersection, vehicle density, and the state of the traffic lights. The circulation of impulses in a network is likewise regulated by the "right of way" of certain stimuli, the network's structure (configuration of the intersection), and its current state (the traffic lights). Which impulses (that is, which information) have preference over others and also where they go is decided by the network's "system of preferences." A network that cannot alter its system of preferences under the pressure of new experiences is unable to learn. A change in the preference rules means that a new conditioned reflex has formed. Feedback either amplifies the links between specific stimuli (the bell always rings when the dog is fed) or weakens them (there is no bell before the dog is fed).

An overload of stimuli causes crowding, hinders the circulation of information, and thereby lowers a network's effectiveness. In left-handed people (whose right brain hemisphere is dominant), the speech center develops in the left hemisphere, and speech-related impulses must run from the left hemisphere to the right and then back again, which leads to "congestion" in the subcortical and commissural pathways.[4] Therefore these people often suffer speech impairments (such as stuttering). When the opposite happens, when there are fewer stimuli than available pathways, "hesitation" occurs, because of the necessity to choose.

A network capable of learning can become conflicted when an old preference system clashes with a new one created in response to new experiences. The simpler the network, the more easily it solves the conflict by an arbitrary choice of which stimuli should have priority. The more complex a network, the more preference systems it contains, and therefore the more possibilities exist for inner conflict. Because a network assembles its preferences without comparing them in terms of logical consistency, preferences formed at various times in its past may be inconsistent and cause a vicious circle in which trapped stimuli circulate in vain. Such a network may manifest signs of neurosis (e.g., compulsive thoughts). A simple example of conflict: a man from a country where people drive on the right side of the road finds himself in a country where people drive on the left side of the road. A higher level of conflict might be a clash between a scientific worldview and a religious worldview. The question of preferences is a question of values because deciding which signal or which information is more important than another means making a value judgment. In the language of cybernetics, the question of "value" reduces to the question of "sorting," that is, how to differentiate between impulses and direct them to different pathways.

So far we have discussed two kinds of networks. The first are simple ones that have feedback but are incapable of learning. These are found in devices such as an "automatic pilot" or a command center in a self-guiding torpedo. Their preference systems are permanent. Networks of the second kind can learn because they possess memory (an anti-aircraft radar system). All living organisms have memory. Even unicellular organisms can form conditioned reflexes and thus modify their preference systems in response to changes in environmental conditions. Both kinds of networks have feedback between the network proper and its organs of input and output.

The third kind of network contains not only this external feedback but also internal feedback, which makes it capable of symbolic activity, a fundamental condition for the emergence of consciousness. Simpler networks circulate information about their environment or about limited parts of themselves (sensory organs) that are located at a periphery of the network's body proper and can therefore be treated more as parts of the environment, except that they are permanently attached to the network. In more complex, higher-level networks, besides these "primary" signals there are "secondary" signals that circulate "information about information."

A set of potential internal feedback links contains everything of which a given network may become "aware" or "conscious." The secondary signals, the information about information, are "abbreviated" labels (symbols) denoting the states of specific parts of the network. What is abbreviated here are not electric impulses but a particular current state of a set of the network elements. When a given symbol appears in the network, it will be acted upon, and internal feedback puts parts of the network into corresponding states. Consciousness is not symbols but processes that "extract" symbols from the network and put them back in for operating purposes.

The "meaning" of a symbol is the set of signals that are potentially connected to the symbol. Such symbolization allows various processes of the network to be generalized. A symbol can denote a set of primary signals originating in the environment ("a tree") or a set of signals originating in the network itself ("sadness"). The more a network can utilize its previous experience (stored in its memory), the more effective it is. This depends on the utilization range of memory data. Psychologists call this utilization a "transfer" of an acquired skill from one field of activity to another.

An animal can be trained so that when it is shown two black dots, it eats just two pieces of offered food, and three pieces when there are three dots. But it cannot transfer this skill from the visual field to another. It will not react to two sounds of a whistle or two touches of a hand; it will react only after it is trained in the field of the other sense. But a human being, understanding the meaning of the signal "at once," has no problem making that transfer. This ability to symbolize or generalize can be expressed by the formula, "If you perceive n signals through *any* of the five senses, perform n actions." Of course, the formula need not be verbal. Words are attached to the symbolizing process only when the task is so difficult that it requires operating with large sets of signals, both primary and secondary. A simple transfer of an acquired skill (without symbolization) occurs directly through forming links between all sensory fields and the generalized directive, that is, the system of preferences that applies to any kind of stimuli. This system then becomes the network's operating rule until new experiences inactivate it.

A network's "intelligence" is a function of the maximum transfer rate of which the network is capable and not of its size. A network can have an enormous memory and still be unable to utilize it in the pursuit of a goal. In that case, the memory becomes useless ballast. By studying its own behavior, analyzing its past, and assigning symbols to responses, a network can "recognize" its operational rules, whereby it becomes "aware" of them and so can modify them "at will" (by recombining or supplementing their elements or by establishing altogether different directives). This "willed" change takes place quickly, unlike the change of a preference system due to external stimuli, which requires replacing one group of conditioned reflexes with new ones, which always happens in stages. Psychologists usually consider a quick change in a preference system a sign of intelligence, in contrast with the gradual change that is observed during training. But under some conditions a sudden change may take place that is not a result of the network's "conscious" activity based on symbolization. It happens when the network attempts to reach a goal without any plan, by trial and error, and accidentally comes across the right combination of stimuli that, because of the success, is immediately preserved. A monkey, given the difficult task of reaching a banana that lies outside its cage and provided with two sticks inside the cage, each too short for the purpose, finds that inserting one stick into an opening at the end of the other—is a good example.

HYLAS   What is a symbol that is not a word?

PHILONOUS   A network usually works with operational units well before it can use verbal symbols. For a predator, for example, such an operational unit is the entire sequence of behavior that culminates in securing the food (catching the prey).

HYLAS   I see what an "operational unit" means in animal behavior, but what is its equivalent in a network?

PHILONOUS   A sequence of consecutive motor directives that form a higher-level whole, as notes form a melody. It is even called "a kinetic melody of movement." This sequence is projected onto a changing background of external stimuli that continuously affect the network and are systemized by its priority systems. Thus a "kinetic melody" is not a constant, given once and for all, but is constantly shaped by feedback and the stimuli that originate from inside the network and form hierarchical sets—that create the "dynamic spatial map" and the "map of temporal links" between consecutive actions. Even in a highly organized network like the human nervous system, similar units of object action (action performed with tools) appear well before the development of language. A symbol here may be a pose, a gesture, or an entire situation (pantomime). We should keep in mind that the role of language in an individual's life, which is to facilitate adaptation through high levels of generalization and systemization of both internal and external information, is secondary to its primary role, which is to communicate with members of the same species. In this latter sense, language is a system of abbreviated stimuli containing sufficient information for one network to send to another clear "action directives" so that processes in both networks become similar (getting "in the same mood").

HYLAS Why do you use this extremely complicated way of describing things that psychology already knows very well?

PHILONOUS Because I am forced to describe instead of providing mathematical formulas which do not yet exist to express the high-level organizational forms of network processes. Note that any behavior of an organism can be described either in words or by a corresponding neural network (a connection map, a "formal scheme"). Theoretically the two are equivalent. But if we want to describe all possible actions of a network—for example, how it can figure out that different triangles are all triangles—our verbal explanation will become extraordinarily long. The recognition of what makes a triangle a triangle, whether large or small, equilateral or scalene, is a tiny part of the larger issue of the "congruity of geometric forms," which in turn is a tiny part of the still more general issue of shape similarity (visual congruence). Psychology's description of animal behavior works well, because the responses there are relatively simple, but with the kind of tasks we have been discussing this method fails, for the reason that what is happening in a network when n processes take place simultaneously to produce the outcome, that is, the organism's behavior, is a very complex resultant of the whole ensemble. For "visual congruence" to be explained, a description of the connection pattern in the visual center of the brain would probably serve better than any attempt at a verbal definition. It thus appears that the most practical description of an object is not the catalog of all its possible states, for that could take forever to assemble, but *the object itself*—in our case, the network itself. For this reason, the search for a logical definition of visual congruence may be futile. A map of neural connections in the network would be fully adequate and equivalent. This method is completely new, never seen in science before.

HYLAS   How so?

PHILONOUS   Logical reasoning is one of the functions of some networks. We can build models of networks that reason logically. A network whose function is to formulate a relation of the type "If p, then q" is *more complex* than the verbal function "If p, then q," its formal-logic equivalent. In simple cases like this, a logical description (the sentence "If p, then q") is always *simpler* than the equivalent network; in complex cases, it is the opposite: the logical description becomes much more complicated and incomparably longer than what is being described, that is, the network itself. Thus we find ourselves in a curious situation: the simplest logical description of a network is *the network itself*, and *logic begins to turn into*—or accrete with—neurology.

HYLAS   Why neurology?

PHILONOUS   Because neurology is, or at least thus far has been, the study of the neural networks of the brain.

HYLAS   So you think that by the mathematization and mapping of network processes we can find an answer to the question regarding the difference between conscious and unconscious processes?

PHILONOUS   That is the only rigorous way to proceed. I have already mentioned how difficult it is to study the mechanism of "visual congruence." But that is child's play in comparison with the processes of symbolization.

HYLAS   What do we already, and really, know about this visual congruence, that is, about how we recognize shapes, objects, or letters to be the same when they appear in a great variety of sizes and forms and can be affected by perspective, lighting, and so on?

PHILONOUS   I can only offer a hypothesis. The number of fibers in the optic nerve is smaller than the number of elements that the nerve is connecting, that is, the light receptors in the retina and the cells in the visual cortex (*area striata*). Similarly, there are fewer sensory fibers that project to the corresponding cortex analyzers than there are cells (acceptor neurons) in them. So a relatively large amount of information must be transmitted through a relatively small number of channels. How is this possible? There is an analogy with a television set, where we have only a single electron beam, which is so narrow that if it fell on the screen without movement, it would draw a point. The beam moves across the screen at high speed, covering its entire surface in a fraction of a second, zigzagging horizontally through every point of the screen's surface. Because our eye cannot perceive changes occurring in less than one-sixteenth of a second we see the TV picture "all at once."[5] The spatial receptor of the brain "scans" with a circulating "beam" the afferent sensory fields.[6] In this way it is possible to transmit a large amount of information through a relatively small number of channels (to transmit two stimuli simultaneously we need two channels; to transmit two stimuli consecutively one is enough). The amplitude of the rotating "beam" is largest when there are no signals at all. This corresponds to the so-called "alpha" rhythm in an electroencephalogram, the regular, sinusoidal peaks and valleys in the electric potential of the cortex, much as the electron beam of the television set fills the empty screen with white noise when no information is being sent. When a figure appears in the visual field, its spatial elements that remain fixed within one period of the scanning beam (one period of the "alpha" wave) are transformed into a time series, and this way a spatial series of dots (the figure elements) is transmitted as a time series of impulses following one another. Thanks to this mechanism, a one-dimensional channel can transmit, via pulsating signals, an image with an arbitrary spatial complexity.

However, such transmission has its drawbacks. First, the speed of reception is limited by the time of the cycle of the scanning beam. Signals shorter than this period give the illusion of movement, because they fall on different sections of the "alpha" sinusoid. Second, the scanning process requires a constant activity of the cortex, an endless circular wave of processes. This spontaneous activity is detected as the essentially constant "alpha" rhythm of the brain's biocurrents. If a signal lasts less than one-tenth of a second, it will fall into the "insensitivity" phase (similar to the "refraction phase of the nerve," a temporary loss of excitability in a nerve fiber after transmitting an impulse). When perception is taking place, currents with various frequencies are summed, the "alpha" rhythm disappears, and fast-changing ("beta") rhythms appear. This is confirmed by the observation that the time interval between stimulus and response in the visual cortex is not always the same, since the impulse arrives at random: if it arrives just before the scanning beam comes to the cortex reception field, the interval is short; if it arrives a moment after the beam has left, it must "wait" until the beam returns after the entire "round."

HYLAS   But what precisely is this beam? In the television set it is a real stream of electrons in a vacuum tube.

PHILONOUS   Obviously there is no such physical beam in the brain. The point is that the number of neurons is higher than the number of afferent fibers, that is, the channels through which information enters. What circulates in the brain is the pattern of sequential connections between the afferent fibers and the analyzer and the related changes in the excitability threshold. Imagine a man in a circular room who must read instrument displays on the wall. He cannot read all of them at once, so he walks from one instrument to another. He can notice a change in any display's reading only if the change persists longer than the time it takes him to make a circuit of the room.

HYLAS   Who is the reader in the brain?

PHILONOUS   The higher-order processes. They are documented by changes in higher-frequency potentials in the cortex during the perceiving and thinking. But it is difficult to isolate these processes because each of them involves large areas of cerebral cortex. There is "a little bit" of each process everywhere, whereas the more elementary processes tend to concentrate in the region of the sensory analyzers. So far, we have described the impulse transmission to the cortex, but to transform the train of stimuli into a perception requires a number of other processes. When we shine a light into a person's eye, we first notice the jump in potential and a perturbation of the "alpha" rhythm in the ocular center (*area striata*), which then spreads into the surrounding secondary sensory fields of the cortex. This is where the network elements are located that enable visual "recognition" based on "visual congruence." It is an extremely complicated process. Consider the recognition of a hexahedron. We can observe it from various angles and distances. From a single-perspective projection of it, we can easily extract a system of equations that show how it would look when viewed from any other angle. The network stores these equations in its memory. The arriving impulses are then compared with the system stored in the memory, and the moment they coincide, a "resonance" occurs, and the network "sees the hexahedron."

HYLAS   Once you speak about the appearance of a hexahedron, then about some equations. . . . What exactly is stored in the memory?

PHILONOUS   Nothing except the ability to transmit a specific series of impulses. An equivalent of this ability is a set of connections that, when activated, reproduces the series of impulses. The mathematical expression of this is an equation, which can be written down, but logically one is equivalent to the other. As we mentioned when we were talking about "visual congruence."

HYLAS   But can such a serial comparison between afferent impulses and memory data work in reality? Recognition that way would take awfully long.

PHILONOUS   What I described was very much simplified. In reality a small, partial match is already sufficient to initiate the organization of the visual field according to the directives of the memory. The visual memory is active and always tries to "impose" its "concept" on the processes in the *area striata*. Or you might say that it is "guessing" what is seen. This is clearly seen in optical illusions, when the incoming information is insufficient, for example, at dusk. The visual memory, attempting to organize the field of vision, keeps "proposing" to the *area striata* various "possible alternatives," which is why, during an evening walk in the fields, you are startled by a person, which turns out to be a bush. What is taking place in the neural network is not just the simple comparison of impulses but a comparative overlaying of extensive processes with the strong tendency to self-organize into specific dynamic structures. These processes occur rhythmically, in the form of neuron firings, but the electroencephalogram registers only the resultant of all the overlapping biopotentials.

Specific processes have characteristic frequencies. In the brain, some rhythms (such as "alpha") are privileged, which is why stimuli with a certain frequency may resonate with the cortex rhythms and increase their amplitude to such a degree that a person suffers an epileptic attack. Luckily, normal physiological stimuli do not have such periodic character. Before the onset of convulsions, the person experiences emotions, most often unpleasant. This demonstrates a connection between emotion and the frequency of oscillations in the neural network. Because aural stimuli are uniquely capable of affecting the periodic processes in a network, music is important to human beings. During perception or spontaneous thinking, the main processes in a network are usually accompanied by others with higher frequencies, like harmonics. The tendency to form these harmonic frequencies depends on the "subjectively perceived situation" (the emotional mental state). When we stimulate the neural network with impulses that have a frequency of twelve per second, frequencies of twenty-four per second may appear in the frontal lobes and those of six per second in the lateral lobes. If a person's mood changes, the ratio of these harmonics changes too. When the frequency of six per second dominates, a person experiences feelings so unpleasant that they are difficult to endure; at twenty-four per second, calm intellectual analysis of the optical illusion that the stimulus elicits becomes possible. When the subject of the experiment is told not to resist the light impulses, the lower harmonics of six per second begin to rise, and the subject is soon unable to continue the experiment. But if an analytical, intellectual atmosphere is maintained during the experiment (with the same periodic light stimulus), then the high-frequency harmonics increase. The slow rhythm usually appears only in dangerous or threatening situations and is associated with the corresponding feelings. It is important to note that the same stimulus frequencies evoke different emotions in different people, but always the same emotions in the same person. This demonstrates a marked individuality of the network processes which is explained by the historical formation of the network's "personality"—its preference systems, memory data, habits, and so on. One hypothesis on the meaning of these rhythms suggests that the "alpha" rhythm seeks visual perceptions and the "theta" (six per second) "pleasant" feelings.

HYLAS   Do we know the purpose of the harmonic frequencies?

PHILONOUS   In principle, yes. The higher and lower harmonic frequencies establish "long range" connectivity—between distant parts of the cerebral cortex. That is why they appear during thinking, especially hard thinking. They are a component of that process, just as the signals from the optical regions secondary to the *area striata* are a component of visual perception. We cannot say that they "mean" anything or have an equivalent in a person's mental life. If we consider a mental process to be a whole, then the harmonic rhythms are part of that whole. We can record them but do not understand their significance, nor do we know what information or what directives they carry from one part of the brain network to another. Yet experiments might shed light on the meaning of these signals. If we intercept an encrypted message that one part of an enemy army sends to another, and we cannot decipher it, we can send it to its addressee and observe the reaction. The reaction will reveal to us the content of the message. Similarly, researchers are trying to record the "cipher" that one part of the brain uses for communicating with another; then they will send that "text" into the neural network at some other time and observe the network's reaction to it.

HYLAS   Have such experiments been conducted?

PHILONOUS   Not yet. It is extremely difficult to isolate a single "encrypted message" between parts of the brain from the huge number of processes taking place at the same time.

HYLAS   I see an even bigger challenge here: the "brain cipher" must be sent to a particular part of the network—and we do not know the addressee.

PHILONOUS   That is not a problem, because, as far as we know, the parts of a neuronal network are not connected like telephones but rather like radio stations: the impulses of a message are propagated in all directions, but only the addressee of the message can make use of them. The "encryption" itself together with its harmonic structure unequivocally determines not only the content of the message but also its sender and destination.

I think that this should end our excursion into the field of the cybernetically interpreted neurophysiology. The phenomena here are so enormously complex that just listing them would require many extensive, multidisciplinary investigations. Let us proceed to the next characteristic of a synthetically (holistically) considered network: free will and selfhood, issues so fundamental in philosophy. By "free will" we commonly mean the subjective feeling of liberty in action in response to a stimulus or a set of stimuli. In this process it is only the act of making a decision, of choosing a behavior, that is always conscious. Once a decision has been made, the resulting action can unfold automatically. It is enough just to have an intention, and its "initiation" is already a task for certain subsystems of the network, which are grouped into operational units. To pronounce a word, we do not need to consciously move the muscles of the tongue, larynx, and lips. It is enough to make the decision—"to push a button mentally," so to speak—and the articulation takes place on its own. With such hierarchic "centralism" the field of consciousness does not need to control its subordinate network processes rigidly; it intervenes only when it is informed by internal feedback that an action is not unfolding as expected.

"Will" is therefore a combination of decisions with their anticipated (expected) results. These decisions, "proposed" by the network and its memory, represent an act of "comparing" the current situation with past ones (again, not in a literal and static sense but dynamic, as we saw in the processes of cooperation between the primary and secondary visual cortices). A choice is represented by the "initiation" of a specific operational unit when it is injected into the network as a traffic rule (a preference system). It is equivalent with suppressing or blocking all information that clashes with the decision, thus steering the organism's behavior. The decision begins at the moment when the cumulative charge of previous information in the network begins to interfere with the inflow of information that clashes with it. After the act of choice, it serves to prop up the selected preference system against the incoming information. Even the simplest networks manifest thusly defined "will" when they, like an automatic pilot, compensate for whatever diverts them from their goal. However, in such networks, "will" is built in once and for all; complex networks can choose among various preference systems. It is possible for them to behave as a person who, "out of a sense of duty," overcomes every pain, fear, and doubt in the pursuit of the goal.

In what sense is "will" free? First of all, it is free from the immediate pressure of the environment, thanks to countering it with previously accumulated data and personal experiences introduced by feedback for the purpose of making decisions. Without feedback of its past the network's behavior would be solely determined by the present situation and its external pressures. Such a network would be unable to choose an "independent" course and steer "against the flow" of events; it would drift with every current. Only inanimate objects drift, passively yielding to the influences of the environment. The ability to seek goals autonomously and choose and pursue them is considered a value in the "moral" sense. The material premise of such autonomy is made possible by the continued possession of one's past, a past that can be activated as needed. This is why the inviolability of the internal feedback links, understood as the full functionality of the systems of connections with one's memory stores, is so important. If a network becomes overloaded with an excessive inflow of external information and a sudden increase in the rate of learning to accommodate it, the network's past becomes negligible compared with the present. It follows that a network acquires "personality" only when the learning process is spread out in time. When there are too many new experiences at once, a network begins to behave as an inanimate object: its past ceases to play an important role in its decisions. It loses the ability to make choices, and begins to passively drift in the stream of events.

Clearly, critical rate of learning, the amount of information that a network can absorb per unit time without being overwhelmed, is an individual variable that depends mainly on what we termed the network's *wholesomeness*. When the previous life experience becomes unusable, it may lead to an internal functional breakdown unless the network can counter it with robust and resistant preference structures. People in the death camps were in this situation. Forced by circumstances to adapt and having lost the ability to counter it with past experiences, people are capable of committing acts that are, in the terminology of various systems of ethics, "immoral," "sinful," or "inhumane." This leads to the formation of new habits and behavioral patterns that may contradict a network's entire personal history. This is a "pathological learning," in that it occurs at the cost of a network's internal continuity or wholesomeness. Similar phenomena appear, with less intensity, when a network loses its ability to learn freely. Freedom of learning is characterized by the rate of the assimilation of new information, which must be such that a network can preserve, at every stage, its past and its wholesomeness. When the inflow of new information is too large and the network loses its freedom of learning, it becomes incapable of assimilating new experiences and making connections with the previous ones, because of which it is no longer able to respond adequately to changes in the environment. Then we often say that a person "lags behind" or "cannot keep up" with new times.

The set of preference systems that decide a network's behavior constitutes its "personality." "Personality" may remain intact even when, as a result of an external force, a network's effectors are temporarily blocked. Enslaved by external forces, such a network maintains its internal wholesomeness or personality, like a ship that, carried off course by the storm, still has a rudder, or like Hamlet, who, imprisoned in a nutshell, nevertheless "counts himself a king of infinite space."[7]

A network's personality changes a little with each new experience and each new decision. Thanks to its past, the network is not absolutely subordinate to the current, transient situation; thanks to the preserved possibility of further learning, it is not entirely dependent on its past either. Its internal transformations in response to new external needs thus occur through the push and pull between its present and its past. "Internal freedom" manifests itself precisely in this relationship. A network that has formed a rigid preference system and does not confront it with new information exhibits in its behavior "rigor," "fanaticism," and "obstinacy." It too loses its freedom, like the one that is subjected to excessive external force, except that the affecting force comes from within instead of from without. Every network typically passes through a period of optimum learning and assimilation of new information, after which that ability diminishes: there is less memory capacity, and its preference structures that dominated most often tend to ossify. This is the origin of "conservatism."

HYLAS   What about the predictability of the network's behavior? Do you think that having at hand your proposed mathematical network theory and knowing a network's present state (having data on all its past experiences and preference systems), we could predict how it will act in a new situation?

PHILONOUS   Not in the case of a neural network.

HYLAS   Why? Heisenberg again?

PHILONOUS   It is true that incalculable (in the Heisenbergian sense) atomic fluctuations may in some circumstances affect behavior (through a momentous change in transmitability of a particular stimulus), but we need not look here for a quantum cause of this indeterminism. Keep in mind that the factors that influence processes in a neural network are innumerable. The excitability thresholds of synapses, that is, the rate of impulse transmission, depends on, among others, body temperature and blood chemistry, both of which fluctuate constantly. These negligibly small oscillations can together exert a nonnegligible effect on the course of a mental process. A microscopic deviation at the beginning of the path of an impulse might grow significant by the end of the path, which might mean a different decision, an unexpected association, a "spontaneous" reaction. This is impossible to predict. Moreover, a network's decision may depend on the "time when the idea occurred," on an accidental, "random" memory record that suddenly "resurfaces in consciousness," and so on. And some networks can manifest an unexpected violation of their habitual preference system. In some fields a network's "value" may actually depend on such violations, for example in the creative work of an artist, engineer, or musician. But spontaneity understood as the ability to abandon old preferences and create new ones is not the only criterion of creativity. Spontaneity must also serve to maximize the richness of new configurations of internal elements. Unexpected behavior in itself does not make an artist—the person may simply be a crank. Someone with a vast memory may be erudite and nothing more. The breaking of an old preference structure plus "internal richness" plus the ability to organize its internal elements into an entirely new structure in a wholesome way—all three conditions must be met to call a network creative. In any case a new configuration can arise only from the elements that are *already* present in the network. Information is thus acquired in two ways: first, externally and second, through the internal recombination of symbols to make a configuration that never before appeared in the network. "Internal richness," "wholesomeness," and "free will" together manifest "character."

Being "free" in the course of its decision making, a network is responsible for all its actions and for its personality that formed by all past decisions since the network's inception. No decision was fully predetermined, and each could have been different (inter alia, due to "chance"). In each case, the probability of such "chance" is small but nonzero. Because a network consists of a huge number of elements (about 10 billion in the case of a human being),[8] it never happens that all the processes and elements participating in two instances of the same situation are exactly the same. This increases the element of randomness. Tradition holds that an actor is responsible for each of his "free acts," yet absolute freedom does not exist. It is a limiting value and therefore unattainable. What is constant in the network is its past, its personal history. Every decision, every step on its life trajectory, arises from the friction between its history and its present. Its history is the accumulation of all its previous decisions and it makes its individuality. Its present is a choice determined partly by individuality and partly by chance, due to the statistical nature of the network processes. In this sense, every decision contains some randomness and may be less or more predictable but never certain, which makes it free. The longer a network exists, the greater the pressure of its past and the less the freedom of its decisions. Yet its individuality is frozen only in the final experience that is death. Before that, a network is always free in its behavior, albeit in an ever-diminishing degree, and loses this freedom at the moment of death. And this, my friend, is all I have to say about life and death observed from the cybernetic point of view.

HYLAS   You have given me much food for thought. But what about the prospect of cybernetic resurrection that you mentioned?

PHILONOUS I meant not resurrection but the continuation of personal existence after an organism's death.

HYLAS Aren't they the same?

PHILONOUS You will see that they are not, Hylas—but not today.

VI

PHILONOUS   Hello, Hylas. A lovely day, isn't it?

HYLAS   Oh, it's you. I was reading and did not hear you approach.

PHILONOUS   What book has managed to engross you like that?

HYLAS   *Notes from Underground* by Dostoevsky.

PHILONOUS   Just the right reading for someone who is immersing himself in the mysteries of cybernetics.

HYLAS   You think? It has in fact raised many doubts in my mind, and I would like to share them with you. Those complex networks of yours obviously are intelligent, can reason, possess goals and freedom of choice, yet what do they have to do with the world of human desires? We human beings, created as we maintain, by evolution, which has built into us many safeguards and skills, should avoid suffering, use induction, pursue progress, and exercise our abilities to the maximum—but the reality is much more complicated. We deceive ourselves. We take a perverse pleasure in suffering, even in our own. We like to destroy. We make thousands of excuses; are full of tricks, locked doors, secret chambers, and odd corners; fall victim to mind-numbing addictions; become slaves of desire, emptiness, and the dark craving for significance, pretense, and dominance. Can you draw a schematic of a self-deceptive network? A network that sacrifices others to the Moloch of "obligation"? A network that finds pleasure in the inflicting of pain? My constructor friend, can there be networks that are blind and fanatic, networks whose magnificent, complicated structure has no other purpose but to abuse and defile itself and the world? If cybernetics cannot help us here, we had better dismiss it.

PHILONOUS   I see, Hylas, that you have indeed been reading Dostoevsky. I understand that your anger is mixed with sorrow, that you are fighting the despair that tends to overcome us when we ponder the human species. But cybernetics is not the right target for your reproach. It is strange to use the language of physics or engineering to discuss the tragic or noble aspects of human existence and psyche. This has not been done before and may well border on the ridiculous. But I rise to the challenge.

Today we were to talk about the kind of immortality that cybernetics might provide in the future. It is not an immortality that we would want; it may be grotesque, awkward, and disagreeable, but there may be no other. First, let us return to the networks. Consciousness, as you know, is only one of many processes in the brain. What we call conscious is injected by internal feedback into the network's special circuits. But the contents of what becomes conscious only partly depend on the will, that is, on an arbitrarily selected system of preferences, and both consciousness and the preference system are carried by the flow of all cerebral processes, as the Earth's globe is supported by Atlas's shoulders.[1] But not all information is admitted to the gate of consciousness. Some processes are fished from their ocean, others not. Some achieve more than a fair representation in consciousness (with respect to their share in the total of the network processes), others are inhibited and suppressed. Some preference systems can be modified during their admission into consciousness. It is easy to change a food preference when we are given information that the food is harmful; it is not easy to part with a worldview when it is shattered before our eyes by proving it logically inconsistent. A network can "fake"—not only its own contents and proficiency but also its relation to the world. An appropriately "raised" automaton will manifest "irrational faith" and "superstition," will perform symbolic gestures to ward off "bad luck" when it sees a black cat, and will engage in mystical and metaphysical disputes, for its behavior is determined by its past. If we place a twentieth-century newborn in the middle of a Neanderthal tribe, the child will grow up not into an engineer or a pilot but a mammoth-hunting raw-meat eater. So much on the subject of a network's "self-deception."

As for "perversity," this feature may be explained to some degree by the pursuit of a surrogate goal, which we already mentioned. Also, it is the price paid for having great ability and talent. Regardless of the number of safeguards, a network with sufficiently high complexity is subject to "deviations" with varying mechanisms. A human being tends to revolt against the society that formed it. Destructiveness, cruelty, masochism, and other vices were not "planned" by evolution, but neither was our love of beauty, music, and art. Keep in mind that in networks there is nothing but co-oscillating sets of processes that reflect and interpret the world. Certain processes or certain frequencies of biochemical transformations, whose task is, say, to look for congruence between shapes, can unexpectedly, in response to some particular signals, undergo a change which causes a sudden drop in a network's internal imbalance. From the cybernetic perspective, bliss, peace, and satisfaction—all those things that art provides—represent a lowering of this imbalance in the course of processes related to the inflow of such information as music or the sight of mountains in the snow. Unfortunately, other combinations are possible too: a network may also create configurations in which the decrease in imbalance is accompanied, for example, by killing. Rashevsky mathematically predicted which geometric shapes a human observer would consider beautiful.[2] One might design a system of processes, a formal network connectivity pattern, that leads to pleasure derived from destruction. That design would be an explanation, not a justification, of course, because, as we mentioned, a network is basically free in its behavior. I have responded to you not as an engineer but as a very inexperienced apprentice of network science who is no match for Dostoevsky either in language or the ability to make an impression.

HYLAS   You are right that we should not blame cybernetics, and it should not be a target of our reproaches (if we should have any in the first place). But what of the prospect of achieving immortality?

PHILONOUS   Well, the continuation of conscious existence, you already know, is inseparable from the processes occurring in the network. Therefore, it would be impossible to "isolate someone's sorrow" so that we could put it into a glass jar—this is as impossible today as in the most distant future. Because the feeling of sadness results from a constellation of processes taking place in a given system, and to obtain it we would need to create the entire system. Please bear in mind that the system can be built from any material and its energy transformations can occur in a much larger interval of temperatures than those in the human brain. We will seek a solution to the problem of continuation by going through a series of stages or consecutive experiments. In the first experiment, we surgically connect the peripheral nerves of two people. This has already been done on lower animals. It will enable one person to experience what another's sensory organs perceive. So one person will see through the eyes of another if the peripheral part of the optic nerves of the first person is connected to the afferent nerves of the second. The next experiment, far more difficult to realize, is joining the neural paths of two brains through a link, which can be either biological (a bridge of living nerve fibers) or any device that can collect the stimuli running through one brain and pass them on to the respective neural paths of the other.

HYLAS Are you sure that even if this experiment were successful, the other person would make any sense of what he feels? I am afraid that the result might be an impression of total chaos and confusion.

PHILONOUS You are absolutely right. Specific stimuli have a "meaning" only for a given network and even there only for the parts of the network to which they are addressed. Simply injecting a random train of stimuli from one brain into another would surely result in "mental cacophony." This is one of the biggest obstacles on the path to the functional joining of two brains. Yet a brain can cope with procedures that are much more drastic, such as the excision of entire slabs of the cortex or even of a whole hemisphere. This kind of brutal surgery does not inevitably result in the breakdown of mental functions: the brain's ability to restore them, even in networks that have been significantly damaged, is enormous. No doubt experiments of the kind I am describing will be cautious for a long time—they will be performed on animals, whose behavior and reactions after the joining will be carefully monitored. But since there are no fundamental, principial obstacles in sight, success will eventually come.

Initially, two joined brains (joined in one or more neural paths, subcortical associative bundles, etc.) will only interfere with each other. But observations that have been made in neurological clinics point to the next stages of our experiment. We know the symptoms of massive brain damage. In the majority of cases, even the most severe perturbations recede after time and are compensated, provided the irreplaceable regions of the cortex remain intact. Functional recovery sometimes happens spontaneously, but more often it occurs after long, conscious efforts under the guidance of expert trainers. The damaged brain replaces the lost functions by repurposing its other parts. Functions lost due to the damage in one region of the cortex are taken over by cortical regions that previously had little or no role in them. For example, after a loss of muscle proprioception, a person loses the feedback informing him about the position of his limbs, which renders movement, especially walking, impossible. But once he learns how to replace the muscle proprioception with visual cues, far-reaching restitution of kinetic ability ensues.

A nice example of a more subtle transfer to a new mechanism is when a person with brain damage, exhibiting symptoms of motoric aphasia, could not pronounce simple words, like "neighbor": having articulated the "neigh," he could not automatically block the innervation of the muscles of speech and started persevering, mechanically repeating the syllable "neigh." What was for him absolutely impossible when he was supposed to say the word that denotes a person living next door became easy after he was instructed to think first of the sound of a horse and then of the sound of a sheep ("neigh-baa"). The problem was solved because different network mechanisms, serving different purposes, had been activated. Damage to large parts of the brain causes impairment, helplessness, and confusion in a person, but in time a lost function can be learned from scratch, and the symptoms diminish or disappear. Therefore, we can expect that if a brain mechanism is not lost but, on the contrary, added—by attaching one brain to another—after the initial chaos, a new *modus operandi* will develop, process coordination emerges on the adaptive learning path, and after some time the complete functional union of the two brains will be achieved. Obviously success will depend on what is connected to what. Joining lower-level parts of a network, those that only transmit information from the sensory organs (e.g., the optic nerve fibers with *radiatio optica*), will cause less disturbance than joining higher-level systems. Joining systems at the highest level of organization, those that integrate and form consciousness, will elicit the greatest and longest lasting disruption (perhaps even madness), because each of the joined parts operates with its own unique "coding method," in which different frequencies correspond to different symbols, different processes influence one another in different ways, and information synthesis is accomplished by different means. Even so, I believe that a unified functioning of joined brains will be possible. It goes without saying that we should join only equivalent systems, the anatomical and physiological units (fibers, paths, brain regions) that correspond with each other. And we should avoid joining just one part of a brain to an entire other, because then, after the initial period of mutual disturbance, the whole would functionally dominate the part. But when we sever the great commissure in both brains and join left hemisphere with left and right hemisphere with right, we can expect that both brains will eventually fuse into a new functional unit. This functional union will be based on emotions and experiences that we cannot imagine, since the subjective perception will be that of a functionally single brain having two separate bodies joined only by a bridge that carries the nerve impulses between them.

Consequently, "plugging into someone else's consciousness" with the aim of subjectively and directly observing its processes is impossible, because the plugging itself at first severely disrupts consciousness (in both the "plugger" and the "pluggee"), and later, when mutual adaptation has taken place, a unified consciousness arises that is not a mechanical sum of two parts but an entirely new functional unit. Direct observation is therefore out of the question; the only possibility is a "participation" in the other consciousness by becoming its "functional part." It follows that the successful functional union of two brains will amount to an end of their previous individual existences: consciousnesses *A* and *B* disappear, and the new *AB* is qualitatively different from either. This rather gloomy statement forewarns us that some kind or some degree of "personality destruction" cannot be avoided here. And we should realize that this applies for the reverse procedure too: renewed separation of the joined brains back into two would mean the end of that newly emerged functional personality. It would entail a new period of disturbances and the subsequent phase of learning or rather "unlearning" of what had been gained in the joining; the two separate brains resulting from this complicated and risky operation would most probably be different from the original brains A and B, because the intranetwork changes and in-depth process reconstructions could not be mechanically reversed: the new recoveries would take place through internal modifications, done for the second time, owing to which we would end up with brains A_x and B_y, not the original A and B.

The best chance of success should be expected in the case of a unification undertaken in earlier developmental stages, when the processes of network formation are still under way and the plasticity and adaptive abilities of the cortex are at their peak. The brains of children would be the easiest to join, and most definitely the brains of fetuses. First experiments, no doubt, will be conducted on monkey fetuses.

HYLAS   How macabre! But what would be the purpose of such terrifying operations which bring us no gain? To produce monsters? Or to prove the impossibility of the "direct connection into someone else's brain"? And what does this have to do with "cybernetic resurrection"?

PHILONOUS   The proof of impossibility that we obtained was incidental, not the purpose of my argument. The goal we are after is to go beyond the individual boundary of life. You will understand this when I tell you that the next, crucial step is to *graft* or transfer a human mind onto a brain prosthesis.

HYLAS   Oh my! How would that work?

PHILONOUS   Essentially we connect a living brain, that is, a neuronal network, to a network of a different kind—electronic (or electrochemical). Obviously people will first need to learn how to construct networks with a complexity in the order of 10 billion functional elements, which is that of the human brain. A general theory of feedback networks will facilitate this. The grafting itself would be done in a great many consecutive stages.

HYLAS   Why?

PHILONOUS   Because connecting a brain to an electronic network at once would lead to a complete collapse of its processes. The primary circuits of the neuronal network must first be equipped with "shunts," because each of them must be represented in a corresponding circuit in the prosthesis. Attaching all the shunts to all the circuits at the same time would perturb the entire network, and the consequences could be fatal. The brain's network is a closed and integrated functional unit; opening it to an outflow of impulses to another, empty network would be equivalent to shorting it out. I may be overstating the danger, but in an operation of this kind caution is wise. Because the "personality" of the human brain must remain intact, a prudent procedure will be to connect the neuronal network to the prosthesis step by step, region by region, so that the living brain can "functionally absorb" or "assimilate" the electronic network. The objective is for the attached network to take over a significant part of the mental processes in the living brain. The next stage, once that has happened, is to gradually reduce the neuronal network. We are not destroying it but "unplugging" it, as is done, for example, in a lobotomy, where the fibers connecting the frontal lobes with the rest of the brain are severed. If we do this in sufficiently small steps, sequentially unplugging only small regions of the neuronal network and taking care not to act prematurely, that is, before the prosthesis network has overtaken the respective function, our functional unit, the combined neuro-electronic network will assume its functions without any significant disturbance, while those functions will gradually disappear from the neuronal side. Eventually, when the neuronal network has been dispossessed, the electronic network will carry the full burden of mental processes, and we will have transferred a human personality into our prosthesis. The electronic network will now contain all the memories, preference systems, impulse traffic rules, and internal feedbacks that previously constituted the personality of the living brain. This "electronic graft" of a living consciousness will be able to exist for an arbitrarily long period, as the material of the prosthesis is thousands of times more durable than the substance of a living brain. Also, any parts that wear out can be replaced. This is the prospect of "eternal life" in electronic or electrochemical brain prostheses.

HYLAS   Hold on. What about the body, the living organism to which the living brain "belonged"?

PHILONOUS   The problem is significant but not on the technological level: there may be moral opposition to the next, and the last procedure. Having replaced the brain, we need to replace the body too . . .

HYLAS   I see. A prosthesis again?

PHILONOUS   To secure longevity, it seems unavoidable.

HYLAS   So the "immortality" that you offer means transferring a person's mind into an apparatus of dead metal? If I were to take your proposition seriously just for a second (which surely does not come easily), I would never agree to that. To exist forever in the form of a thinking metal cupboard? Maybe you're joking, Philonous.

PHILONOUS   I am rarely more serious than I am now, my friend. The totality of mental processes can be excised, extracted, separated from the short-lived, impermanent living body only through its slow transfer to another substrate that will endure.

HYLAS   All right, if we put aside for a moment the moral objections here, what is the guarantee that a mind transferred from a living neuronal network to a bunch of metal wires will not be deformed, mutilated, and dehumanized? Can this be considered with any seriousness at all? The prospect is ridiculous, insane—a world in which people are replaced by metal boxes equipped with electronic sensory organs . . .

PHILONOUS   You were supposed to withhold the emotional judgment of the issue for a moment, if I understood correctly. My task was to show you the only real, or at least probable (as of today), path to immortality in the future, not to make value judgments about that path.

HYLAS   Fine. Still, what is the guarantee that this procedure, even when done as gradually as you say, will not damage the living brain (wires stuck into its living tissue?) and turn it into something nonhuman?

PHILONOUS   The procedure need not be bloody at all. Replacing the 10 billion neurons of the cerebral cortex with vacuum tubes is impossible, of course. Even with transistors, solid-state devices, which are 90 percent smaller and 90 percent more efficient in energy consumption, it would be impossible. To support the operation of an apparatus equivalent to a brain, about 100 million watts would be needed, whereas a living brain uses barely 100 watts and thus is a million times more efficient—as well as almost a million times smaller than a hypothetical "solid-state brain." On the other hand, owing to the enormous difference in the speed of signal propagation between the electrical and nerve impulses, the thought processes in a crystalline brain would be about 100,000 times faster.

HYLAS   Are you saying that during the "grafting" operation a person would be chained for years to a giant machine-building?

PHILONOUS   Von Neumann has calculated that in theory an artificial brain could use 100 *billion* times less power than it does now. The first step on the path to improving its efficiency, from a vacuum tube to a transistor, has already been made. This step will undoubtedly be followed by others. An artificial brain of the future will certainly be smaller and more efficient; theoretical limit even allows for artificial brains that are hundreds of times smaller than the human brain, though possessing an equivalent number of functional elements. Science, then, considers it possible to bound "Hamlet's personality" in a nutshell.

HYLAS   But what about that awful "grafting" procedure itself? It sounds like vivisection.

PHILONOUS   Today it is thought that the number of central neuronal groups, that is, closed circuits in the brain that play an essential role in the emergence of consciousness, does not exceed 10,000. Each group contains a number of circuits, closed loops of impulse circulation, which were discovered by the brilliant researcher Lorente de Nó.[3] It is entirely possible that the functional joining of a neuronal network with a nonneuronal network can be accomplished without the subject's discomfort. Keep in mind that such an operation will not be available for another thousand years; by then, medicine, neurophysiology, and neurosurgery will have the means at their disposal. Also, this operation is in one respect not that different from what occurs in a living brain all the time: its building material is regularly replaced through metabolism. Except that our replacement of material substrate is much more radical—from protein-based to nonproteinaceous. In any case, continuity and integrity of the transferred processes should be preserved in every step of the transfer.

HYLAS   Even if everything goes smoothly, the idea of "thinking metal boxes" as the next stage of human development is unacceptable to me. But we might avoid this macabre vision by transferring the mental processes of one living brain to another, equally neuronal, protein-based, and alive, only created synthetically. What do you think of that?

PHILONOUS   I see no principial impossibility there, but, paradoxical as this may sound, it would be much more difficult to do than grafting a mind onto a nonliving prosthesis. Constructing an entirely passive receptacle without any trace of memory and "personality disposition" would be straightforward compared to creating an artificial, fully developed, alive, but at the same time "empty" brain. And there is another important issue: a person's new living brain would begin to experience, soon after the operation, various ailments and defects, and would quickly come to the end of its existence.

HYLAS   Why is that?

PHILONOUS   Every network has a limit in "information capacity," which includes both its memory and the total amount of information, from outside and inside, that can circulate in it. Experimental and clinical data indicate that the human brain is not far from this limit, particularly as it ages. (This is one of the reasons why older people cannot remember recent events but have no trouble remembering the distant past.) In an overloaded brain, even a small hormonal perturbation that lowers the neurons' excitation threshold just a little could totally block the transmission of impulses, causing insanity, personality disintegration, irreversible damage. Not long ago a substance was discovered that appears to inhibit synaptic excitability, and when administered to healthy people, causes symptoms of schizophrenia. The substance was isolated from the blood of schizophrenics.[4] So we may conclude that after the mental processes of an old man have been transferred to a new brain, it will be close to the limit of "information capacity" and be able to function for only a relatively short time.

HYLAS   Very well, but what about the electronic brain prosthesis?

PHILONOUS   We can build one with additional "stores" or "functional reserves." But as you can guess, that still does not promise any kind of "immortality," since only an infinitely large and infinitely complex brain would be able to store an infinite (or at least enormous) number of memories, not to mention anything else.

HYLAS   So this whole project of "transferring" a mind is a fantasy?

PHILONOUS   No. Nothing in science rules it out. As I have already said, using electric current and improved functional equivalents of neurons, we should be able to build a network that is ten or a hundred times more capable than the human one.

HYLAS   Great. Creating a "synthetic genius" then?

PHILONOUS   A general theory of networks, once we have it, will enable us to construct networks with whatever characteristics we like, as long as they are allowed by the laws of nature. The great English mathematician Turing provided a theoretical proof of a network that can "do anything that is possible."[5] Thus in the future it will be able to construct a network that can compose a symphony or figure out all possible paths of evolution on other planets.

HYLAS   Philonous, you're laughing at me!

PHILONOUS   What, is your human dignity offended? If you are not bothered by the sight of a crane 10,000 times stronger than you, why resent a machine 1,000 times smarter than you? As an *energy* machine augments human power, so an *information* machine augments human knowledge! Scientific progress makes us face more and more difficult problems. Twentieth-century mathematics is far more complicated and requires much more mental effort than tenth-century mathematics, yet our brains today are the same as they were in the year 1000, because evolution in mathematics is a million times faster than the evolution of the human brain (i.e., its structure and function). If we cannot lift a weight, we build a machine that can. If we cannot solve a theoretical problem, we build a machine that can. Where is the insult to human dignity here? After all, if we ever put together a "synthetic Einstein," we have put together it, not it us!

HYLAS   I guess it is the sense of the *superfluousness of human beings that concerns me*. A "synthetic genius" does not need us, our cooperation, or our control the way cranes and steam hammers do.

PHILONOUS   What is the problem as long as they work *for us*?

HYLAS   So you believe that someday machines will exceed human beings in all respects?

PHILONOUS   Someday? It is already happening, Hylas. Every electronic calculator solves problems that the best mathematician cannot in a lifetime. There is a silly and naive myth about factories of the future as bright halls full of automata among potted palm trees, with a man in a white coat at the central console supervising the production. But it is nonsense. Take a chemical plant today, in which reactions in hot gases occur at lightning speeds. To harvest a valuable compound that appears in a gas stream for a fraction of a second, one needs to maintain its source reactions. The steering and control must be on a timescale of milliseconds, which is impossible for a human being to do because our nerve impulses are not quick enough. The man in the white coat therefore has nothing to do in the factory; the production is run by an electronic brain.

HYLAS   But when that brain breaks down, he is there to fix it.

PHILONOUS   Another electronic brain, connected for that purpose, will fix it faster.

HYLAS   And when the other brain breaks down too?

PHILONOUS   And when the man falls ill? There is no *regressus ad infinitum* here, just a hierarchy of automata that mutually control themselves, a closed circle. Obviously anything may break down. Today it is people who fix; tomorrow it will be machines.

HYLAS   Yet your argument shows that so far an electronic brain beats a human being only in speed.

PHILONOUS   True. But let us consider an example where the issue is not speed but a higher-level, integrating characteristic of the network. As you know, a thought process grows more difficult the more elements (notions) need to be taken into account at the same time. It is easy to perform elementary arithmetic operations from memory, but difficult to calculate the fourth root of a number with ten figures. Yet it is just an issue of "short-term memory," that is, keeping track of the results of each partial step we make in the mathematical reasoning where the operation instructions (multiply, store the result, then divide, etc.) are fixed from beginning to end. But when a task is one of generalizing many facts into a theory, in the course of the generalization process each consecutive stage also modifies the instructions, which are not predetermined and fixed but are the outcomes of consecutive transformations. If we attempt to develop a theory of gravity that is more general than Newton's from the data of astronomy, physics, and mathematics, such a huge number of factors must be considered *at the same time*, that only an extraordinarily capable network can manage that. Einstein was in possession of such a network. Of course, not everyone can be an Einstein, but in the future everyone will have at hand a thinking machine with unlimited abilities.

HYLAS   This does not bode well. For a while, increasingly powerful electronic brains will work on tasks that people can still understand, at least roughly. But then a gap will open and widen: our thinking machines will provide us with solutions to problems, solutions that we will be able to use but not understand. Automata will spread their dominance until people shrink to the level of thoughtless servants and begin worshiping the iron geniuses like gods . . .

PHILONOUS   Just think, my friend, your prophecy regarding the human species has already come true, and in the far past at that.

HYLAS   What are you saying? I don't follow.

PHILONOUS   The emergence of electronic brains has indeed started the evolution of the tools for artificial thinking. And machines can potentially gain independence from the human race, just as other consequences of human social and manufacturing activity did in the past. Take division of labor, which resulted from the emergence of society, or novel tools and methods of production: all of this created machinery that, having gained independence from the human will, started to influence the lives of individuals so much that sometimes this machine—the state—has become an object of worship. This analogy is neither accidental nor superficial, Hylas. People ought not, now or ever, lose control over the work of their hands and brains. They ought not surrender to placid thoughtlessness, intellectual laziness, and rosy optimism, eager to believe that this or that invention or this or that social organization automatically guarantees the coming of the golden age. No indignant "Man is still the crown of creation" will change the facts—and the rise of ever better electronic brains, which no one can banish from our lives once they have entered it, is an undeniable fact. Unless people consider carefully all—the good, the bad, and even the worst—consequences of the development of electronic brains, the computer evolution may be more ravaging than crises, economic catastrophes, joblessness, and the chaos of the capitalistic free market. For this very reason we are talking so much about cybernetics, trying, often in vain, to understand what it has to say about phenomena that are apparently so distant from one another, such as evolutionary biology and psychology, or the general theory of information and sociology.

HYLAS   Don't forget eschatology, the science of final things, since its subject is life eternal and therefore includes your grafting of a living human brain onto an inanimate prosthesis. Do you believe that people will ever attempt to realize such a transfer of the psyche from a living human body to the dead metal of a machine?

PHILONOUS   The system of privileged rules of thinking, that is, the system of preferences, applies not only to individual neural networks, my friend, but also to societies. In this sense culture is a system of historically formed preferences that channel people's responses to external and internal stimuli. Today we are well aware of the conventional (that is, history-dependent) and therefore relative nature of most ethical norms, moral imperatives, and established rules. The idea of transferring a living human psyche into a dead machine appears to violate a number of our fundamental habits of thought; it appears humiliating, improper, inhumane, and unacceptable. But we cannot rule out a profound shift in our norms and preference systems in the future, after which our view of this issue may drastically change. Keep in mind that we are talking about an operation that will become possible only thousands of years from now. The evolution of electronic brains is hiding in its bosom many powerful challenges to contemporary worldviews. Suppose electronic brains equal to us in intelligence are brought up as religious believers or even bigots. Can you see how terrifying opponents would they be for all religions? What sophistries would theologians have to spin when confronted with the manifestation of the "spirit" in electrical wires and vacuum tubes? However, a problem far more significant and difficult to solve than the conflict between spirituality and cybernetics is, how should people spend their time in a society where absolutely all the production of goods is automated? In an exclusively consumption-oriented, passive society, how will people who live in great material luxury and great mental sloth face the fact that every human activity will be rendered absurd by the availability of its superior actualization by thinking machines? These are the problems for the human mind to address! You are demanding from me answers that I do not have, Hylas. In our history it has always been that the unknown—occasionally even a product of our hands—appears first, that questions are raised first, and only later, in sweat and labor, answers emerge. Often spread over many generations, the answers are imperfect or partial, but while the problem is being clarified and solved, new unknowns and new question marks rise on the horizon. Let us end today's discourse with this. We have just one more problem to discuss, but it is most complicated: sociology cybernetically understood.

VII

HYLAS   Philonous, I think I have discovered the fundamental difference between an electronic brain and an organism. When an electronic network solves a problem, it can immediately forget it. In this way calculators are totally unaware of the mathematical operations they have just done. But a living brain never forgets its past, or at least an outline of it. So for a neuronal network the whole life of the organism from beginning to end is like a single task; the network builds its personality, character, and individuality in the course of solving it. Therefore, such a network cannot start "truly anew," make itself empty and blank again as an electronic network can. Right?

PHILONOUS   This is true for currently existing electronic brains but will not be true for those constructed in the future. I have already mentioned that network engineers are not interested in imitating the total of a brain's activities; they only want to make devices that can mimic a specific, narrow subset of the nervous system's functions. They always intended to build not an "autonomous" network "that can form a personality" but only steering and controlling systems for industry, machines that can reason logically or shoot down an airplane, and the analogies with the functions of the nervous systems that emerged in those devices, surprised them too. Only recently they turned to machines that can imitate the behavior of living organisms, that is, machines that can learn based on conditioned reflexes and manifest elementary tropisms, for example.

The irreversibility of processes in a neuronal network is closely related to its complexity and to some extent its building material, as physicochemical changes occurring in it (for example in the course of development or aging) substantially affect its function. It is especially interesting to note that the human brain's network forms during development, and not only functionally. Large parts of an infant's brain are almost totally inactive; they gradually "plug into" the functioning network between the second and seventh, or even tenth, year of life. This process manifests anatomically as the myelinization of the nerve fibers in various parts of the brain, mainly in the frontal lobes, which, being the site of higher-level mental processes, become functional latest. This is no doubt related to some features of neuronal network's functioning, one of which is the subjective perception of the passage of time: an hour is considerably longer for a child than for an adult. This difference is not an illusion but a result of an increasing differentiation and complication of the network, along with the processes in it, with age. The irreversibility of the processes in a neuronal network thus has numerous causes because such a network develops both functionally and structurally, whereby it increases its informational capacity. Of course, we could create an electronic network that "develops" in the same way. Initially it would have a relatively simple "active nucleus," but in the passage of time and as new needs appear, various supporting subsystems would attach, obviously in a way that allows no conflict between the original nucleus and the subsystems. But let us now leave this aspect and turn to our main topic, which is the application of cybernetics to studies of the structure and function of society. Society, as a system (organized set) of elements interconnected by feedback, is, paradoxically, more like an electronic brain than like a living organism.

HYLAS   I fail to see this similarity.

PHILONOUS   An electronic brain can approach a new task with no memory of what it did before. A neuronal network cannot regroup its internal elements in a way that would return it to its original position as society can. An organism and a society share some features—in both there is circulation of information, matter, and energy, and both are subject to the fundamental laws of cybernetics (e.g., regarding the measurability of information, feedback, and preference systems)—but there is a principial difference between them: society, because of the looser connection among its elements, possesses a degree of "internal recombination freedom" that no living being does. That is why society is not an analogue of an organism, and it is a mistake to draw any sociobiological parallels between the two.

HYLAS   But an electronic brain, you say . . . ?

PHILONOUS   Well, I may have slightly exaggerated the similarity between an electronic brain and a social structure, because society differs, and quite significantly, from the existing electronic brains. The possibility of constructing an electronic network that would be a functionally equivalent model of a society does exist, at least in theory. But we are closer to constructing a network equivalent to the human brain.

HYLAS   Why? Society must be structurally less complex than a neuronal network consisting of 10 billion elements.

PHILONOUS   Except that every element of a society (i.e., a person) is itself a neuronal network, hence a society possesses a vast range of possible responses, its complexity equals to the "personal" one raised to an exponent that is the number of members of the society.

HYLAS   Now the problem seems hopeless. What is the solution? Are you saying that in order to study societal processes we would have to build an electronic model of monstrous size?

PHILONOUS   No monster brain would be needed. A phenomenon known from physics and biology comes to our rescue: the laws of statistics.

HYLAS   I hate hearing about the use of mathematics right at the beginning. It takes us into a thicket of abstractions, in which we get completely lost.

PHILONOUS   The mathematics necessary for the creation of a theory of social processes is indeed thorny, but we will not go deep into it, especially because this field is still incomplete and imperfect. And for the same reason we cannot succeed in forming a unified theoretical model of human social activity. At best we will only shed a little light on certain aspects of this immense task. Let us start with the basics.

The feedback links that an organism possesses are principially *negative* only: they always act to *reduce the influence* of whatever diverts the network from its goal. An antiaircraft gun reduces its aiming error in each consecutive shot. An automatic pilot reduces the airplane's deviation from the set course. Negative feedback in a network subject to diverting forces is therefore characterized by a series of diminishing oscillations (error to one side—correction—error to the other side—correction—hit). But there is also positive feedback, which amplifies instead of reducing a stimulus. Some devices in radio engineering, such as so-called reactive amplifiers, are built on this principle. In an organism, however, positive feedback appears only in pathological states, because it works to the detriment of the network (and the organism).

HYLAS   I recall something you said about an increase in electric potential in the cerebral cortex in response to light at certain frequencies, which can lead to an oscillation that causes an epileptic attack. Is this an example of harmful positive feedback in the cortex?

PHILONOUS   Unfortunately, it is not so simple. It is not clear if that feedback is positive or negative.

HYLAS   How come?

PHILONOUS   The feedback does not respond to the stimulus immediately but always with a *delay*. Time is required for the impulse to enter the network, for a circuit to close, and for the correcting impulse (response) to be emitted. An autopilot in an airplane responds to a wind-caused deviation from the set course by forcing a deviation to the opposite side, but always with a delay. The two deviations (one caused by the wind, one caused by the autopilot) overlay and cancel each other so that the airplane stays on course. Except that in reality there are small deviations from a straight line: first a deflection caused by the wind and then another, to the opposite direction, due to the pilot's reaction. If, however, the wind does not blow continuously but comes in gusts at intervals equal to the delay in the pilot's response, the deviations to the left and right repeat in a way that causes the system to oscillate. This tendency to oscillate is the Achilles' heel of all self-regulating systems with negative feedback. In general, when a stimulus is not constant but acts with a specific period and this period approaches the characteristic delay of the system, the system begins to oscillate. Because feedback, attempting to compensate for a stimulus-caused deviation, provides the corrective deviation with a delay, the result, in the case of a periodic stimulus, is just the opposite of what was intended: instead of decreasing, the oscillation amplifies to the point of exceeding the limit of the system's mechanical stability. Every feedback system falls into oscillations when the delay of its reaction equals to one half of the period of the external stimulus.

HYLAS   I'm not sure I follow.

PHILONOUS   Imagine that our airplane is pushed off course by gusts of wind with the same period as the delay of the pilot's response. The result will be a series of repeating deviations in opposite directions such that the flight path becomes a sinusoid. Different systems have different delays. In electronic networks, they are thousandths of a second; in neuronal networks, tenths of a second. The mechanism of the whole phenomenon can be shown in this figure:[1]

delay of reaction to the stimulus = A
stimulus   _____
reaction   - - - - - - - - - - - - - - - -

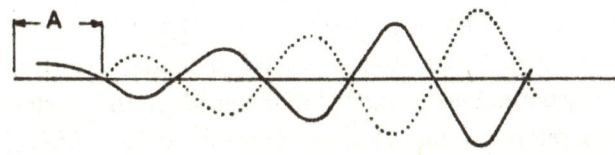

If the correction is greater than the stimulus-induced deviation, an increase in oscillation results. When the correction and deviation are equal, we get an oscillation with a constant amplitude. In the case of positive feedback instead, the relation between stimulus and response differs in that the correction is *to the same side* as the deviation:

stimulus   _____
reaction   - - - - - - - - - - - - - - - -

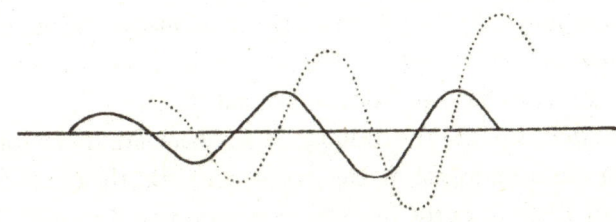

Thus, even though negative and positive feedback enact correction with the opposite and the same sign as the deviation, respectively, both cases may lead to sustaining or even increasing the oscillation instead of dampening it. Yet it is worth noting that negative feedback quenches all stimuli except those that act with a period equal to the system's reaction delay (or its multiple), whereas positive feedback never quenches any oscillation.

HYLAS   I see, sort of, but don't know why you are talking at such length about induced oscillations in feedback systems.

PHILONOUS   Because such oscillations lie at the heart of many important social phenomena. For example, in a capitalist economy, the alternating periods of boom and bust.

HYLAS   What type of feedback is in operation there, negative or positive?

PHILONOUS   Both. Suppose a producer supplies a product that brings profit. To increase the profit, the producer supplies more of the product. The market absorbs it. You have positive feedback between supply and demand, hence an increase in production. But the reaction delay, that is, the delay in providing more of the product, begins to affect the market. When supply exceeds demand, you have overproduction. The market does not absorb the product, the producer profits less, and therefore produces less and lays off workers. After the boom, the bust begins. A steep enough depression causes a crisis. A crisis occurs when the existing social structure fails as a result of a sharp increase in economic oscillation. If, in the worst of the depression, the social structure changes (e.g., through a revolution), the structural factors that induced the oscillation disappear. If there is no change, the system returns to relative equilibrium after a while, and the whole game starts over. But besides positive feedback, negative (corrective) feedback also operates in society, in the form of human intervention aimed at stopping or at least decreasing the oscillation through appropriate regulatory action. We will get to that later. Let us now take a look at the general characteristics of a social system.

First, the delays in the functioning of feedback are much longer in society than in a neuronal or electronic network: they could be on the order of many months or even years compared to milliseconds. Second, and this is essential, the operational rules of a social system are nonlinear. A system is linear if changes within it are proportional to their causes (the responses to the stimuli). The rules describing such a system have the form of differential equations and can be mathematically described with high precision. In contrast, methods for studying nonlinear systems are much more complex and do not provide entirely certain results.

HYLAS   Why is that?

PHILONOUS   Trying not to get too much off course, let us limit ourselves to defining the main difference between the two types of systems. A society or a neuronal network of the brain is a nonlinear system; a group of atoms or stars is a linear system. Linear systems have laws that are constant, so when we discover them, we can predict future states of a system based on knowing its present one. The laws of celestial mechanics enable us to predict with great accuracy the positions of planets and stars in a hundred thousand or million years. Nonlinear systems also possess laws, but they are not permanent and change over time. In the social system, people do not necessarily behave the same way when put twice in the same situation because of their changing mental state, which makes the latter an essential process parameter. Being aware of the consequences of an atomic weapon, for example, may explain why, in a situation no different from one that in the past always led to war, people refrain from engaging in an armed conflict today.

HYLAS   But people did not have atomic weapons in the past, which hardly makes this a valid analogy.

PHILONOUS   You are right; it was not the best example. Let us put the idea this way: if a nation manifested great bravery but little restraint many times in the past and the result was tremendous loss of life, in a situation that previously provoked violent action, it might now show restraint.

If the universe were a nonlinear system, it would not have constants like the speed of light,[2] the Planck constant, the atomic constants, and so on. Yet the universe may actually be a nonlinear system, and all the values that we consider constant are subject to change, although we are unable to observe that because it takes place in intervals spanning hundreds of billions or trillions of years. Which would also explain why all our cosmological hypotheses of what the universe was like in a remote past or what it will be like in a distant future have so little credibility and certainty. In the realm of physical laws we can dismiss what is just probable and irrelevant for predictions spanning thousands or millions of years. (A diagram of a nonlinear trajectory will be a curve. Very short segments of a curve can be treated, with an approximation justified by practice, as a straight line, which means that within a limited time interval a system can be considered linear and its laws constant.) However, in the realm of social systems and their laws, which are nonlinear, we cannot dismiss that.

One of the largest and oldest nonlinear systems that we can examine is the entire evolution of life on our planet. I already spoke of the cyclical links among its elements and of the feedback relations that operate in a single evolving species. But I have not mentioned the feedback between different species of animals and plants. Such phenomena are studied by population dynamics. Take two animal species, one a carnivorous predator feeding on the other, herbivorous. There is periodic oscillation in the number of individuals in both species, because the predators reproduce faster only when they find enough prey. When the carnivores become so numerous that they eat more herbivores than are born, the number of carnivores decreases because of the scarcity of food. After some time (the feedback reaction delay), this decrease causes an increase in the number of the herbivores which are now less hunted, and the cycle starts again. Here we have a linear relation, since in a time interval of, say, a few hundred years, deviations from the direct proportionality that result from the evolution of forms (i.e., the phylogenetic changes that affect population dynamics) can be safely ignored. This is why the differential equations of Volterra, who studied these phenomena, describe the dynamics of such a population so well.[3] Yet in reality the matter is much more complicated, because feedback does not connect only two species but includes all the animals and plants living in that particular environment. This feedback, moreover, is responsible for phenomena on a global scale. For example, consider that the total mass of living organisms on our planet more or less equals the total mass of free oxygen in the atmosphere.

For us, it is significant that in the course of evolution a dynamic equilibrium was reached among all existing species through this feedback such that the population did not fluctuate far from the average.

Only human intervention can drastically disturb this equilibrium. For example, an unexpected but experimentally confirmed result of Volterra's equations is that protecting the herbivore species also increases the number of individuals in the carnivore species. Similarly, the application of an insecticide against plant pests can, via feedback, affect the existing biological equilibrium so that severe disturbances of population dynamics arise, which may even lead to the extinction of a species that the insecticide does not harm at all. Ignorance of specific feedback connections may therefore pervert the human intention in nature and also, as we shall see in a moment, in social systems.

Oscillations in social systems are fundamentally different from those in biological populations. First, feedback connections in biological populations are relatively constant, given the very slow rate of change in the environment and the organisms themselves (in the speed of signal transmission in their neuronal networks, response types, lifespan, type of food, etc.). In particular, *the delay* in response remains the same over millions of years. The operation of feedback in society, on the other hand, has accelerated during human development, because of organizational efforts (labor) and the consequent changes in the methods of production and communication. Second, animals are subject to the dynamic laws of the system to which they belong but they cannot knowingly influence those laws. In society, human beings can. The difference is like that between the way animals and people "arm" themselves—the former by developing tusks and horns and the latter through engineering.

HYLAS   Where are you headed with all these evolutionary divagations?

PHILONOUS   To point out that a biological species evolves not only because of the selection pressures of climate and geology but also, and often predominantly, because of feedback of the kind that Volterra described. A newly appearing species may disturb the biological equilibrium of others and change the feedback loops in such a way that that it causes extinction of a species that it does not directly predate. Precisely this type of phenomenon was responsible, according to current views, for the mass extinction of the giant Mesozoic saurians.[4]

A biological population that is internally unstable, with increasing oscillations in one parameter and decreasing oscillations in another (because feedback does more than just regulate the *number* of individuals!), evolves spontaneously through interspecies feedback, whereas an internally unstable and oscillating social system may neither break down nor evolve structurally into another, new system. I am not comparing the two whole sets of the dynamic laws—of biological populations and social systems—but simply highlighting this essential difference between them.

When we do compare the dynamic laws of a biological population and of a social system, we find that the former can exist for a long time in a stable form only if it achieved equilibrium and is internally stable. In contrast, the latter can exist for a long time even when it is unstable—for the reason that a social system can be "forced" into stability using coercion (force). This is why changes in social systems throughout history tend to have a violent character, unlike in biological evolution, whose course never changes abruptly.

Social systems exhibit oscillations in a number of parameters but economic oscillations are always primary and political and cultural oscillations secondary. These secondary, induced oscillations, in turn, affect the primary oscillations through changes in communal psychological attitudes and people's behavior—another demonstration of the cyclic, coupled nature of these phenomena. Oscillations in all social systems formed throughout history always tend to increase in amplitude. Perturbations growing more and more severe typically leads to the destruction of the system by revolutionary forces that oppose the effort to save and preserve the existing structure in an unchanged form.

Three methods have been employed to suppress oscillations in an unstable social system. The first two use force in a system that has been basically unchanged; the third destroys the existing system and creates a new social structure by recombining its elements either impulsively or according to a preconceived theoretical plan. As a rule, any such plan presumes the new system to be linear or at least close to linear. We are going to describe briefly each method. The first uses feedback with "excessive correction." When the negative phase of an oscillation induces pressure "from below," from the masses, this method responds with greater force "from above," from the authorities. In a capitalist system, we label this protective response to oscillations as fascistization of the society.

If the elements of a social structure were objects and not people, the only reason to criticize this method would be its technological primitivism, like that of an energy-producing machine that uses some of the energy it produces to dampen its self-induced oscillations, which results in less energy that can be used for other purposes. In this analogy usable energy represents social activity whose aim is to satisfy people's needs, while the energy used for dampening the oscillations corresponds to activities that serve not to meet people's needs but to preserve the existing social structure. Because the elements of the system are people, fascistization means not just a mere waste of social energy, but more importantly, a violation of personal integrity in the name of preserving the integrity of the social structure. But the constituents of the vast social system network are the neuronal networks of individual people, and so to keep the higher-level network stable, personal freedom and the developmental potential of individuals are sacrificed. As we know, this method, i.e., the use of force, turns people, neuronal networks, thinking and autonomous units, into passive, mechanical elements, which is the worst thing that can happen to a network-type system.

In practice, the use of force to stabilize a capitalist structure is not revealed to a society as such. Rather, a metaphysical doctrine is created to conceal the real goal behind many spurious ones, which justifies the actions taken. The doctrinal goal may be single—external expansion needed to address an alleged lack (e.g., of a "living space")—or they may be multiple—doctrines based on discrimination and segregation of the members of the society into "the better" and "the worse," etc. The arguments range from pseudoscientific to utterly irrational (such as the nation as a mythical union of blood and land), but the purpose is always to get people to accept the imposed situation. From the cybernetic point of view, this is a case of having surrogate goals replace real ones, so it is a social pathology analogous to the pathology of learning that we encountered in neural networks.

The second method for damping oscillations in a capitalist system is based on planned changes in the feedback delays. It originated from studies by many economists, including Keynes.[5]

HYLAS   Are you saying that these economists already used methods of cybernetics to solve economic problems?

PHILONOUS Cybernetics was not yet born at the time of Keynes's school, but yes, in a sense. Keynes created quite a complicated theory, but I will discuss only one of its elements, which is a good example of an attempt to save capitalism by changing parameters of feedback links in the system.

As you know, in nineteenth-century capitalism Marx discovered the law of pauperization of the proletariat and accumulation of capital. This process seemed to have a linear character, and Marx predicted that a series of deepening crises would lead to the eventual collapse of the system. But the process was linear only in a certain time interval, and an intervention changed its characteristics. I have already mentioned the feedback in supply and demand, which causes overproduction, worker layoffs, and the resulting decrease in a market's purchasing power; the feedback between an increase in unemployment and a decrease in demand thus precipitates a crisis. In the case of a neuronal network, you may recall, we spoke of such divergent oscillation as a "short circuit" or epileptic seizure.

HYLAS So crises are the epilepsy of capitalism?

PHILONOUS *Cum grano salis*, one can say that. Keynes claimed that the investment rate depends on the profit that producers expect, and those expectations in turn depend on the market situation. Thus the market situation depends on the investment rate, and the investment rate depends on the market situation. Other variables are at play, but at the given capital concentration and production methods the variables that we pointed out play the decisive role. Therefore, Keynes advises that time shifts in elements of the economic processes, particularly the investment rate, which is the easiest to control, should be planned in long-range projections. This changes the parameters of the existing feedback. When a society's purchasing power decreases, the "investment emergency stores" should open up, which would, at least to some extent, damp the oscillation amplitude. Such a policy of continual intervention assumes the existence of intervening organs. For example, the state, guided by economists who study oscillations in market parameters, might dedicate a part of the budget to large contracts (investments). Note that such investments are made not to meet societal needs but to lessen the threat of an oscillation (a drop in purchasing power).

HYLAS   But an organism also possesses "interventional emergency stores," in the form of its network's negative feedback, which compensates for all harmful influences of the environment (this is the basis of the regulating of temperature, blood pressure, the chemical composition of the blood, metabolic rates, etc.).

PHILONOUS   This is the big difference between a society and an organism. In an organism, self-induced oscillations are abnormal, pathological; in a capitalist system, they are natural, inevitable—all we can do is use "buffering" or "moderating" tools, of which "investment emergency stores" is one. The economic situation of a society is a resultant of a large number of factors linked by feedback. Economists' forecasts and state interventions work to regulate this situation, to keep it in relative balance. Any change in production methods or in the environment (world markets) can disturb it.

HYLAS   How real is this threat?

PHILONOUS   The real danger now is the "second industrial revolution," the mass automation of production processes, which has begun in our time. Automation leads to a drop in the price of the final product, because automata work faster and are cheaper than human beings. The interest of the consumer here coincides with that of the producer: the former wants to buy cheaper, the latter wants to produce cheaper. But automation also leads to unemployment, which in turn reduces a society's purchasing power. For this reason, automation has so far been conducted in the United States with caution and below capacity (barely 10 percent of factories have been automated, and less than 6 percent of the total capital has been invested in automation), to keep the negative feedback from becoming positive. Yet a society in which the means of production remain private property cannot achieve full or even major automation, since owning an automated factory makes no sense: the producer cannot expect a profit if nobody works in the factory and therefore nobody earns the money needed to buy the product. Thus automation undermines the system's fundaments—private investment and the circulation of money and goods. Capitalism has not yet reached the critical stage of automation, but it will happen within two decades. At present, investment returns coincide to some extent with societal interests, but a change in the system of production will make the two diverge. Progress in technology threatens the system's stability. American economists are working hard to solve this problem but have not yet succeeded.

HYLAS   Indeed, the future of capitalism does not look rosy. But at present you consider it stable enough, owing to its control over the economic oscillations, right?

PHILONOUS   An organism is an original, primary entity, irreducible to anything else, whereas a society is a secondary phenomenon. The interests, so to speak, of an organism's parts must be subordinate to its existence as a whole; it would be nonsense to say that a leg or a lung is more important than the body it belongs to. But the subordination to society of an individual's interests should happen only to the extent necessary for the individual's benefit, for his freedom and well-being. Of course, the term "necessary" can be filled with various contents. Here we enter the difficult realm of normative systems, which do not require justification. What are or should be the goals of an individual? What is society allowed or not allowed to do to an individual? For what purpose does a society exist? In considering questions like these, we step beyond the bounds of both cybernetics and sociology. We are asking not what is taking place in a society but rather what should be taking place so that its members can be happy. Cybernetics, like every other science, is silent about human happiness. But once we have established, on the basis of a free choice, what constitutes the fulfillment of human needs and the broadest limits for the fullest development of individual freedoms, we can address cybernetic sociology with the following questions: Does this particular societal structure guarantee people the given rights and the given number of degrees of freedom, and can this structure even be realized? Are its dynamic laws linear over a long time or are they nonlinear and therefore predictable only in approximation? Will the structure be internally stable or not? Will it show developmental tendencies toward the loss of equilibrium, toward a reduction in degrees of freedom, or toward self-induced oscillations? Should it be protected against detrimental changes in its parameters? If yes, then how? It may turn out that the ideal structure cannot be realized; or maybe it can, but its developmental characteristics will have serious drawbacks; or the structure will be stable for, say, a century, and then will succumb to degenerative tendencies. And after long deliberation we may choose another structure that does not meet all our criteria but includes greater possibilities for development and guarantees the emergence of such long-term societal automatisms and linear processes that will spontaneously steer it toward states with increasing numbers of degrees of freedom.

I hope it is now clear that capitalism cannot be this structure, not only for moral reasons (which demand condemnation of the exploitation that is inherent in it), but also because of objective characteristics of its internal dynamics, which hinder technological development toward full automation. Capitalism's use of economic means to dampen oscillations may be less costly in human terms than the physical violence used by fascism, but the consequences are still negative in all spheres of life. The work of von Neumann showed that the dynamics of social processes have the character of a game (in the formal, mathematical sense).[6] Everybody who lives, and chooses to do so, in a capitalist system must accept its rules of the game, which are ruthless. No one asks you if you agree with those rules or know how to use them. Whoever does not know or refuses to follow them must be destroyed. Economic perturbations and oscillations in market feedback decide human fates. Everything that has value, by virtue of that fact, becomes goods. A ceaseless economic war is raging, and the stronger people, not the better, are the victors. A prosperous capitalist country like the United States is similar to a heat engine in that it can function only if a difference in temperature exists. The heat in the boiler by itself cannot do anything until it is conducted to a place with a lower temperature. Therefore, every heat engine must have a cooler or condenser; for the States the "cooler"—or heat sink—is foreign markets and colonial countries. Capitalism requires and maintains unequal development, because, as we noted, what is best for society is not always profitable. We could expand this critique by showing the relation between the authority of the state and economic processes, proving the secondary nature of that authority and demonstrating how negligible effect it has on the general direction of social life. But none of that can alter the negative judgment of this system expounded in the works of Marxist sociologists. Therefore, let us turn instead to the search for the ideal social structure. It is not a simple task, and we lack the knowledge to arrive at any concrete, detailed model. Even so, building such a system is within human possibilities—important in the twentieth century, a time of great hopes and great disappointments.

HYLAS An ideal society being one that meets all our requirements—which we must define first, on the basis of our beliefs and worldview? And we must decide on an acceptable range of individual freedom, yes? Hence: freedom to act, without diminishing the same freedom of other members of the society; freedom to develop personally, to be oneself, and to exercise one's talents; the maximum possible fulfillment of all life's needs. All of this, naturally, regardless of one's origin, birth, race, or nationality.

PHILONOUS It is hard to disagree with you, and yet—no matter how paradoxical it may sound in view of what was just said—I would begin our discussion of the "ideal system" not from the individual but from the society, or perhaps from both sides simultaneously. Instead of adopting your focus on personal freedom, let us consider all mutual interactions among people simultaneously. Human abilities and characteristics complement each other not only in the economic sphere but in all spheres: creative work in the sciences and arts, family life, friendship, and love. The degree of this mutual complementarity tells us about the stability of social links; the degree of personal freedom tells us about the developmental ability of a society. The highest level of both—that is the formula for our model. Such a society develops not by simplifying the existing links and its structure, not by subjugating its members, but on the contrary, by increasing the complexity of its structure. The more information circulates in a network, be the network neuronal or electronic, the more activity fields, and the greater variety of needs, talents, occupations, and tastes it has. It is precisely this differentiating and fragmenting dynamics of the ideal model that counteracts the emergence of oversized and ossifying institutions, which regularly appear when human groups organize on the principles of hierarchy with features of socially harmful automatisms (in contrast with beneficial automatisms, of which we will speak later). Any individual ambition must find a societally organized outlet and a path to its maximal realization, possibly transforming societally harmful tendencies into beneficial ones along the way. Therefore, an individual's growing responsibility for himself and his fate must be accompanied with an increased feeling of connection and complementarity with others. Only a structure that takes into account the interests of both the whole and its parts can ensure both the free development of individuals and the growth in adaptability of the society, so that the society can change in response to its environment, whether the environment is local, global, or even interstellar. This is how I see the ideal model, my friend.

HYLAS   But this is not a model, just a list of criteria that you submit to cybernetic sociology, which will determine whether or not this set of parameters can coexist in the same structure. What is the next step?

PHILONOUS   True, my proposal does not define parameters as measurable quantities, which cybernetic sociology needs if it is to find, out of thousands of possible variants, the model that best meets the criteria. We are very far from such an undertaking, as we are very far from assembling a blueprint for the proposed model, since cybernetic sociology itself is still only a set of postulates and observations, not a fully developed branch of science that can be applied.

HYLAS   Are there fundamental laws that govern a society?

PHILONOUS   One can call any law fundamental, just as one can say that human nature is *naturaliter christiana* or *naturaliter socialistica*.[7] But this is arbitrary, a fetishization of laws, and does not bring anything new to the table.

HYLAS   Marxists say that the fundamental law of capitalism is the pursuit of maximum profit. Are they wrong?

PHILONOUS   No, but that characterization of the system is insufficient. Observe the form of a statement offered as an objective social law. "Pursuit of maximum profit" clearly points to purposeful activity and *implicite* assumes there is feedback in the system. "Pursuit of the maximum satisfaction of people's needs" also assumes there is feedback. Laws of social systems differ from those of material systems, such as collections of atoms or stars, in that social laws must include feedback and therefore have a purposeful character (even when no one is aware of the purpose), whereas material laws do not. All social systems thus belong to a class of networks that have feedback and are capable of purposeful action and learning (see the rise of national cultures).

HYLAS And what does a cybernetic analysis of a socialist system look like?

PHILONOUS All social systems in history arose spontaneously, but socialism, an attempt to construct a society based on known laws, did not. Marxist historical materialism defines a fundamental rule common to every social system: the formation and organization of interpersonal relations are dependent on the method of producing goods. This law is as universal as the laws of thermodynamics. Just as a machine that obeys the laws of thermodynamics may be efficient or not, capable of economical and long-lasting performance or not, a social system that obeys the Marxist law of dependence between the two relations may be internally stable or not. A necessary condition for building our new system is the socialization of the means of production, because private ownership, as we have seen, gives rise to socially harmful economic oscillations with repeating periods of unemployment and subordinates individual lives to the economic law of value. Moreover, in the long run private ownership does not permit automation and consequently blocks progress in the methods of production. But if this socialization is a necessary condition, it is not sufficient to guarantee the natural, noncoerced stability of our new system. There are many possible ways to organize socialistic production, and not every one of them sets into motion the social automatisms that would ensure success. Therefore, experimentation is critical for choosing the right model.

HYLAS What do you mean by noncoerced stability? And what are the "social automatisms" that you mention?

PHILONOUS Any system exhibiting divergent self-induced oscillation, such as capitalism, can be made stable by the use of force. Without force, the oscillation, increasing, will bring the system down, like an unbalanced machine that comes apart because of runaway vibration.

The force applied can be physical—in fascism, for example. It can also be economic—capitalism uses that kind of pressure (the effect of the law of value) to keep the social structure stable, but there can be a physical component as well (in dealing with labor unrest). The "ideal" system should, naturally, refrain from the use of any force.

Except that the construction of our new system must begin within an old system and it has many unprecedented features. The first is the necessity to overcome the resistance of those who wish to maintain the old system. No other construction work takes place under such conditions. That resistance cannot be overcome without force. The second feature is that the constructors of the new system cannot remain outside of what they are building, unlike the constructors of conventional machines. The boundary between the constructed and the constructing disappears in this process. The third feature is the use of people as construction elements, which makes the constructors' actions subject to moral evaluation, which in general is not the case in other construction projects. The fourth and final feature is the simultaneous operation within the system of two kinds of laws: established and objective. In conventional machines, only objective laws operate. Both kinds of law have feedback, but the principial difference is that an established law can be violated without affecting the whole system, whereas with an objective law this is not possible. Therefore, an established law carries with it penalties, which an objective law does not need. The relations between the two are complicated. The functioning of an objective law can be changed by an established law if and only if the established law brings about a structural change in the system (e.g., a vote on the nationalization of industry and land turns a capitalist system into a socialist one).

HYLAS I don't understand. How can any established law violate the causal dependence of interpersonal relations on production relations?

PHILONOUS Yes, this fundamental law of society is universal, as we said. But every machine, besides being subject to the laws of thermodynamics (e.g., we always take from it *less* energy than we put into it), also exhibits many regularities specific to its operation. And the dynamic rules of a social system correspond to these operational rules of a particular machine. If this machine does not exist, its operational rules obviously cannot manifest. Only when we build it, according to a plan, the rules may begin to manifest themselves.

HYLAS   Why do you say "law" sometimes and "rule" sometimes?

PHILONOUS   This differentiation stems from what we might call "constructor empiricism." If we knew absolutely all the objective laws in a certain field—e.g., the laws that govern atomic transformations—then we would not need to resort to long and troublesome experimentation but could deduce from the totality of these laws the best blueprint of the thing we want to build (e.g., a nuclear reactor or engine). In practice, however, our knowledge is never complete. Einstein's theory of gravitation explains facts that the previous Newtonian theory could not, but it does not explain *all* the facts. In the future there will emerge a theory of gravitation that corresponds even better with what we observe in the real world. And so on, forever. Important for a constructor are not only the laws he already knows but also the laws he does not yet know or, what comes to the same, unforeseen consequences of the laws that are already known in the field. When airplanes were built on the knowledge of aerodynamics, the engineers could not avoid problems caused by the unforeseen induced oscillations in the structures. In general, every constructed device exhibits regularities in its operation, some of which are foreseen, and in fact desired, by the constructor, and some of which result from laws that are not yet known or are known but were not taken into account. A theorist who studies collections of atoms or stars knows that his predictions based on the application of a law in its current form will not be matched exactly by what really happens. A constructor cannot accept that; instead, he uses trial and error, empirically studies all the regularities exhibited by the device under construction, to arrive, through the elimination of what doesn't work, at the project realization that will be satisfactory. Knowledge of the general laws of a system, which in practice is always incomplete, is therefore insufficient to unequivocally guide a constructor in his work. One can say that a constructor acts with incomplete knowledge of the objective laws that govern the device he is constructing. And if this statement applies in some extent to all areas of technological effort, it does so especially to the area of sociological construction, where our knowledge is still relatively meager. This is the first reason for making a distinction between laws and regularities-rules.

The second reason, also empirical, is the difficulty of deducing the specific operational rules of a device from general objective laws. Every refrigerator is subject to the laws of thermodynamics, and yet those who build refrigerators deal very little with the abstract laws of thermodynamics; they pay much more attention to the technical properties of cooling devices, that is, to the rules of their operation. A detailed consideration of the scientific problem that we are discussing here would require time and extensive research, neither of which is at our disposal. For our purposes, the two differentiating aspects that we mentioned should be sufficient. Thus the universal sociological laws apply to all social systems, but those systems also exhibit in their functioning specific regularities dependent on their structure.

If the flywheel of a steam engine is incorrectly engineered, and therefore unbalanced, and the machine breaks into pieces due to centrifugal forces, it is a manifestation of the laws of mechanics but at the same time a consequence of the given construction and the effects that are specific to it (e.g., the resonance vibration of some of its parts), and we call this a regularity or a rule that all the engines built from the same blueprint will manifest. The imperfection that induced the vibration can be removed by making changes or improvements in the blueprint, which obviously does not change the universal objective laws governing the machine. In the same way, we can remove certain negative phenomena that manifest in the functioning of a social system by changing its structure, which can be achieved with the help of an established rule. The point is that the established rule should address the true and objective causes, the systemic source of the problem, not merely *mask* the observed perturbation.

HYLAS   Mask?

PHILONOUS   Meaning to cover up, camouflage. An established rule that operates, that is, is obligatory, in a system can make it difficult to discover the objective dynamic law of the system.

HYLAS   You have lost me.

PHILONOUS   Imagine a traveler on a ship on the ocean. There is a storm, the ship rocks, the traveler becomes seasick, which is a manifestation of objective rules of physiology: the motion stimulation of his cochlear labyrinth causes, via a reflex pathway, cramping of the stomach with familiar secondary results. But at the last moment the traveler suddenly learns that the ultimate manifestation of seasickness on this ship is punished by death (for such is the rule established by the captain). With tremendous exertion of willpower our traveler will refrain from "feeding the fish." We have here a suspension, at least apparently, of an objective law by an established rule, without any change in the system. Speaking more generally, people may become seriously ill but, under the threat of punishment, manage to prevent the illness from manifesting itself. In this way authorities can mask the effects of certain objective regularities in social dynamics.

HYLAS I see now. Yet your traveler might be unable to hold back the manifestation of his seasickness. But that is probably beside the point . . .

PHILONOUS This is precisely the point, my friend. You have put your finger on it. The objective laws of organisms, both biological and social, or, even more generally, the operational rules of systems with feedback, are not strictly deterministic but *statistical*, and that is the reason they sometimes seem to be violated. The traveler will *probably* but not certainly keep from "feeding the fish." As for the statistical nature of the rules of social systems, a summation of a large number of individual processes results in a significant regularity with only rare exceptions, so we can make predictions of how individuals will act. We may indeed come across a capitalist who, moved by charity or mental illness, donates his factory to the workers, but we can safely rule out that suddenly all capitalists will offer their factories to the proletariat, thereby transforming the social system. Likewise a few individual molecules in a pot of cold water may move with the speed that corresponds to the temperature of boiling, but it is statistically impossible that by chance all the molecules will suddenly move at the same time with that speed, making the water in the pot boil without our supplying any heat.

HYLAS And what about the social automatisms that you have mentioned?

PHILONOUS This topic requires a more detailed analysis of our model. In the capitalist model, the essential independence of the economic processes from the governing processes is evident. The former, dominating the latter, determine the path of the society's development. Governing links connect the administrative center with the periphery, that is, the society. In contrast, economic links (of production and product circulation, sale of goods) have no center and are always peripheral. It is precisely in economic links, i.e., in the connections between the parameters of supply and demand, that the automatisms of capitalism operate. They appear wherever an individual's personal interest coincides with societal needs. For example, it is in the producer's personal interest to react to increased demand by increasing the supply. Yet it is just one part of a much broader issue. A social dynamics automatism manifests in an equilibrium that is established between needs and their fulfillment in all spheres. All societally needed professional positions get filled, even though capitalism principially lacks specialized organs to do this filling. It is the result of the constant "pressure" of economic conditions, which can be compared to the biological "pressure," that developmental expansiveness that, operating in the processes of organismal evolution, causes life to be present wherever the conditions are right. Or, to give another example, just as we automatically achieve saturation by pouring a salt solution over a layer of solid salt, the saturation of societal needs is achieved by creating an army of reserve workers. The "pressure" of economic conditions means imposing on individuals the necessity of making a living as their personal motivation. In our new model, this motivation should be replaced by a different one—the awareness of the usefulness of one's labor to society. Admittedly, this is the weakest point in our theorizing, because we are assuming that a person who appreciates the societal benefit of his efforts will work as much and as well as he can.

Production methods today, along with the sharp division of labor, have the consequence that an ordinary worker sees only a very small part of the production cycle. People whose work encompasses a whole cycle—artists, scientists, craftsmen—are the exception. Every worker thus contributes just one drop in the ocean of the society's work and it is extraordinarily difficult for him to track his contribution to the universal production. The less coerced the dynamic continuity of a social system, the more the interests of the individual, that is, the subjective reason for his activity, will coincide with the interests of the society. The establishment of spontaneous and stable motivation, free from individual perturbations and without the propping factors of personal economic involvement, has been rendered practically impossible by the modern division of labor, which overabstracts individual contributions into the total of society's production. Our new socialist system makes various compromises to create social mechanisms that substitute for that subjective motivation. But data show that after almost forty years of practice a worker's effectiveness in capitalism is still higher than that in our socialist system. It obviously follows that the vector of personal motivation, even with those specially designed social mechanisms, does not coincide with the vector of societal needs in socialism as closely as in capitalism.

HYLAS Are you saying that the data prove the superiority of capitalism over socialism?

PHILONOUS I would not jump to that conclusion. The efficiency of current gas turbines is lower than that of a combustion engine, yet experts agree that the future belongs to jet propulsion. The point is that the existing constructions need to be improved. But before talking about improvements, let us take a closer look at the mechanism of these phenomena.

The studied system has both the centralization of power and the centralization of production control, which is why we name it centralistic. During its operation, it manifests regularities that the plan has foreseen, such as the lack of a reserve army of labor and an increase in real wages, but also regularities that the plan has not foreseen, in particular certain long- and midperiod oscillations. *Oscillations in production* may be hidden, showing not in the amount of the product but in its quality. They result from an overlay of the oscillating production plans (an initial scarcity followed by an increased effort to fill the gap) and the oscillating market supply (temporary, periodic gaps in distribution).

HYLAS Why do we have these oscillations?

PHILONOUS   There are many reasons. The structure of the system shows a tendency to "move the decision-making up the ladder." The place where a decision must be made, that is, a response to a piece of information (e.g., about the market supply/demand ratio) is formulated, is pushed higher and higher in the hierarchy of power.

HYLAS   What causes this strange phenomenon?

PHILONOUS   It is an objective rule caused, first, by perturbations in individual motivation and second, by the institutional nature of the organs that control production. Ideally a production plan should take into account the current societal needs but the future needs as well (for example, production of the means of production). In theory these should be two sides of the same coin, but in practice they are not, due to the various length of the feedback links. Hierarchical institutions may serve the purpose for which they were created but also exhibit dynamic regularities of their own, which are manifested in their tendencies toward subordination of individuals and autonomization. The institutions exhibit conservatism and tend to grow and ossify by adhering to the same mode of operation once it has been established. By virtue of being a part of an institutional structure, the institution's members turn into links in information transmission, which decreases their autonomy in decision making. The institution represents the feedback system between demand and supply, but these links are very long in comparison with the numerous peripheral automatisms, which in capitalism consist of producers and consumers whose personal interests shorten the supply-demand connection. In a centralistic system, the links that regulate production are as long as the links of governing, and every link must pass through the center.

HYLAS   I've heard somewhere that institutionalism has been blamed for the faulty functioning of the socialist model, but I have to say that I am not convinced. Capitalism also has large institutions—monopolies and trusts, for example. It also has public service organizations, which often are large, but they still work with exceptional efficiency and are able to fully respond to societal needs.

PHILONOUS   Precisely: they respond to societal needs. Please, note that in capitalism, the constant pressure and the incessant regulatory function of those needs keep the institutions, above all the public service ones, from degeneration and autonomization. In reality, speaking of institutions in general is like speaking of the complexity of neuronal networks in general: it disregards the specific purposes that those institutions (or networks) serve. You already know that excessive complexity can be detrimental to a network, and it also applies to an institution. In evolution, the regulatory factor is natural selection, a selective function of feedback that eliminates every biological construct that does not serve the preservation of the species. In capitalism, an institution that does not bring a profit (e.g., an overbureaucratized travel agency) automatically disappears because it goes bankrupt, just like a species that loses in the battle for survival goes extinct. In the centralist model, institutions are not subject to that criterion or pressure by real, objective conditions. They continue growing in size and complexity.

HYLAS   Why?

PHILONOUS   Over time, in the relatively small group of those who govern, there emerges such a concentration of production-regulating feedback links that the group's "information capacity" exceeds its limit, which necessitates a further expansion of the central apparatus of governing. Such a system is like an organism that, deprived of its automatisms, i.e., reflex centers, must consciously and with focused attention regulate the beating of its heart, the pressure and chemistry of its blood, its breathing, the processes of tissue metabolism, and so on. Such an organism would be unable to do anything besides maintaining its own life processes in a semblance of equilibrium.

Centralization, excessively increasing the density of feedback links, blocks (or at least impedes) information transmission and at the same time lengthens its pathways. Instead of direct connections between supply and demand, hierarchically layered "relay stations" appear in the system, and the delay between stimulus and response increases. And we have already explained the role of a delay in a feedback system. In capitalism, the delay in production processes, that is, the time interval between a change in demand and the resulting change in supply, has a crucial effect on the oscillations; in the socialist model, the delay caused by the lengthening of feedback pathways (periphery—center—periphery) is the most significant.

When a delay between stimulus and response is similar to the time interval in which the stimulus acts, it becomes an essential parameter of the system, that is, it begins to actively affect the processes in the system. An example is a film: the impression of movement that the audience has is due to the frequency of the stimuli (frames on the screen) approaching the delay in the reaction of the neuronal networks of the audience. A similar phenomenon in a social system is responsible for phase shifts in the production cycles of factories that cooperate in making the final product; if there is no buffering reserve of intermediate products, the result will be dissociation and desynchronization of production. This causes factory downtimes and depressively affects the powerless workers, which only further decreases labor efficiency. A kind of vicious circle has been created.

HYLAS Wait. I have just realized that "centralization" exists in an organism too, since reflex centers are subordinated to the nervous system. Also, doesn't a social automatism exist in the capitalist system, which through the functional anarchy of its elements gives rise to oscillations?

PHILONOUS Yes, it was a simplification when I spoke about an organism. Functions that an organism performs as a whole (e.g., the search for food) are indeed always governed by its "centralized" nervous system. These are mainly processes that govern the relation of the organism to its environment. But in the area of internal processes, the reflex (vegetative) centers and local muscle automatisms play the crucial role, and they operate on the basis of feedback between cells and tissues. The loss of connection between any part of the organism and its central nervous system does not lead to a local disintegration of function or manifestation of tissue antagonisms—precisely thanks to this local correlation. Only a disintegration of the local feedback links, which shows in the form of unregulated, unhindered growth, that is, neoplasia (cancer), poses a threat to the organism. Thus it is a consequence of feedback disintegration not on the highest level, i.e., in the central nervous system, but on the lowest, peripheral level. As you can see, even organisms that have been evolving for billions of years are not completely safe from a breakdown in an inner functional correlation. In this light, the enormous difficulties that the constructors of social systems face become more understandable. A hierarchical institution's tendencies toward an unlimited growth and an emancipation from the regulating effects of the social organism is indeed analogous to tissue cancer, but I do not think that analogies like this have any cognitive value, given the differences between the two types of system. In particular, the two have different goals, and their construction elements are not the same: while the elements of a social structure—people—represent autonomous values as individuals, the elements of an organismal structure have value only with regard to the organism.

Capitalist automatisms do regulate social dynamics ad hoc by influencing individual motivations for action, but at the same time, through feedback links, they can cause long-term disturbances in social dynamics. Here I am neither praising nor criticizing, only describing.

But let us return to the topic. We have discussed how oscillations are induced by the concentrating and lengthening of a system's feedback links and by the weakening of the individual motivation to act. The authorities counteract the dangers of not reaching a production goal, of neglecting some production areas for others, of lowering labor efficiency, and other related phenomena, all of which ultimately result in lowering product quality and quantity, by putting into action a special "administrative-persuasion" apparatus. It persuades people to do things and thus takes place of internal motivation. This is why every increase in production, whether it be sowing, skimming, or harvesting, requires a battle cry and a campaign to create the impression that a universal and extraordinary effort is made for the good of the society. As social automatisms and subjective motivation fade, the authority changes from an organ that plans and regulates into an organ that intervenes in every production cycle or even in every area of social and cultural life and through a barrage of instructions, promises, slogans, directives, and prohibitions exerts administrative pressure on its citizens. The people engaged in this activity are not producers but supervisors of the production. They form a structure similar to the hierarchical pyramid of the machinery of bureaucratic administration. The necessity to constantly focus the general effort and attention on the issues of production turns the means for achieving the goal of satisfying the needs into an autonomous goal.

A capitalist producer whose goods do not satisfy societal needs endangers his existence. This danger automatically informs all his efforts, so that ultimately his personal interests will coincide with the demands of the market. A socialist producer is supposed to meet the target of a plan, and the judgment on whether it satisfies societal needs is not his. His meeting the target does not always satisfy societal needs, and when it doesn't, the feedback link between supply and demand is weakened again, only this time the weakening is not central (in the bureaucratic apparatus) but peripheral (in the factories).

The construction plan does not foresee the phenomena described above, and thus instead of studying and analyzing them as dynamic rules of the existing structure, it systematically disregards them. What the plan in a centralist model does foresee is the equivalent of the "system of stimulus preferences" in a neuronal network, but only the information consistent with the plan is permitted to circulate freely in the feedback system that the authorities maintain between themselves and society (press, radio, official releases, etc.). However, as we know, an organism will change an old system of impulse preferences in response to new experiences, otherwise it would not be able to survive. In the centralist model, the authorities expend a great deal of effort and means to keep the original preference system, that is, the original production plan, intact. They ignore the rules of social dynamics that were unforeseen and yet became evident. Whatever is not consistent with their preference system is blocked and sequestered beyond the reach of the feedback that operates between the authorities and society.

HYLAS   What exactly is blocked?

PHILONOUS   Various kinds of information: people's dissatisfaction with the scarcity of certain goods and also dissatisfaction with the official (i.e., by the authorities) demand to express *satisfaction because that is what the plan did foresee*; certain results of scientific research (note that cybernetics itself was off-limits for a while); news of the detrimental effects of the overgrowth and functional changes in the governing machinery (bureaucratic and repressive); news of perturbations in agricultural production, and so on. The greater the gap between actual information and what is postulated by the plan, the more effort is needed to mitigate the societal consequences of the phenomenon. Let us take a simple example. A person turns on a steam engine without the Watt governor (which is a simple self-regulating device based on feedback) because he is certain that the engine will work without any problems. But in the absence of the *automatic* regulation of its operation, the engine speeds up until vibrations threaten its whole structure. The person at first tries to ignore it, pretends that "everything is running smoothly," and claims that the engine is all right. But seeing the increasing vibrations and aware of the danger, he begins to reinforce the machine with iron bars. When that doesn't help, he decreases the steam intake. But now the engine is running too slow for its purpose, so he must increase the intake again. The closing and opening of the valve repeats periodically. The perturbations of the engine's operation and the self-induced vibrations—these are the oscillations in the production parameters of the centralist model, and the person's reaction is the reaction of the authorities of the model. The system, then, also manifests *oscillations of the second kind*, in the political hesitations: the periodic broadening and narrowing of the limits to action and thought in governing, producing, science, culture, and so on. This vicious circle results from self-induced societal oscillations on the one hand and the delayed and hesitant response of the authorities on the other. When the cumulative effect of many phenomena that were unforeseen by the plan reaches the limit and "exceeds the excitability threshold" of the authorities, they intervene to stop or at least diminish the oscillations ("deviations") by any means at their disposal. But because they address the consequences and not the causes, their intervention only has a temporary effect, yet it seriously affects psychological responses in people who are the system's elements.

It is remarkable that so far no theorist has attempted to find in the periodicity of these oscillations an objective rule that would be a consequence of the system's dynamic structure itself. Instead, explanations of "distortions" and "deviations" have always employed a subjective-psychological terminology: in the "tightening" phase, "dogmatism," "doctrinism," or "commandeering" operate, in the phase of "release," "oversensitivity" or "petit bourgeois attitude"; the avoidance of decision-making by pushing it up the bureaucratic ladder results from "securantism" or "comfortism," a fear of disrupting the production, and the "bureaucrats' callousness"; noting negative phenomena in life is "slandering" and pessimism, and so on. But it is not surprising after all. These are words of a specific language whose purpose is not to explain phenomena in a scientific manner but justify or interpret them in a way that is consistent with the plan. As the gap between the societal facts and the a priori claims of the original plan widens, the apparatus of this ad hoc constructed language must grow as well.

Remember the story about the tribe living on the plain that I told you in our third discussion? Their discovery that distant objects disappeared beyond the horizon was a consequence of the Earth's sphericity. But the people who disagreed with this scientific explanation had to find another—for example, magic: the distant objects were snatched by "unclean powers." In the socialist model, such an "unclean power," responsible for everything that doesn't work, is "the remnant of capitalism in people's minds." This creation of a language that falsifies the objective character of phenomena in the system is a typical psychological response of the people who must live in it. As a consequence, the scientific plan for the organization of a new society slowly transmogrifies into a set of dogmas that, like a religious faith, cannot be disproved by experience. Another psychological consequence is that decent and subjectively honest people turn into cruel tyrants.

HYLAS This is strange indeed. Are you saying that the cause of this is the objective rules of a social system? How is that possible?

PHILONOUS As a rule, the people who commit various transgressions, abuses, or even crimes in the name of the authorities were once, before doing injustice to others, revolutionaries who spent years fighting for justice with heroism, determination, and loyalty to the idea, the characteristics that we usually do not associate with tyrants known from history. Don't forget that the creation of a new system requires the use of force to overcome the resistance of the expropriated classes. The theoretical plan foresees this violence but assumes that, as the new system strengthens, the need for it will go away. The plan does not foresee any oscillations, which, as it turns out, also can be stopped by force. At the beginning, the distinction is clear: repression is used against the enemies of the new, better system. If there are any oscillations in this initial phase, they are negligible. No force is required to deal with them; often persuasion will suffice.

HYLAS   Persuasion?

PHILONOUS   The elements of our construction are people, not inanimate parts in a machine. This we should not forget. The captain of the ship with the seasick traveler in the example I gave you earlier says to the traveler when the waves are still not that rough, "Sailing is wonderful. You will love it in no time. The rocking, once you become accustomed to it, is quite pleasant. And we will soon arrive at a marvelous port. Suppress your malaise, please, and rejoice in the prospect of our future!" Eventually such appeals will have no effect. When in the face of all directives and prohibitions, the oscillations increase, the authorities will not restrict themselves to persuasion, appeals, and incentives, and focus all efforts on those parameters that changed most threateningly. This "mobilization" improves the values of those parameters, but addressing the symptoms instead of the causes unintentionally activates new feedback connections or amplifies those that have not been evident. After some time, this leads to new perturbations. And they grow. This is the time when the first step is usually taken, the first reshuffling in the machinery of power, by which repression of enemies of the system imperceptibly transitions to repression of friends. The coercion apparatus was already in place—this is very important. A small change in the direction of its application is all that is necessary.

Suppressing the enemy's resistance is an unavoidable necessity. Enemies operate in various ways; why can't they also cause the drop in agricultural production or labor efficiency? Hence the distribution of regulations and decrees begins with an intended purpose of removing harmful symptoms from social life. Established laws begin to mask objective laws. On the violently rocking ship even the threat of death cannot stop the sick man from "feeding the fish." What is left then? Only one option: to gag his mouth. Simultaneously a vocabulary is created to justify each consecutive step on the path to coercion. In this way, the machinery for defending the new construction from enemies slowly turns into machinery for maintaining stability at the expense of freedoms and friends. This is a sequence of actions by which every ruler becomes a tyrant. The new language leaves no doubt. Voices of dissatisfaction and manifestation of protest are labeled as voices of enemies and instigators and calls for changes as calls for the return of the old system of social injustice. Step by step, this can lead to the worst crimes against humanity committed in the new system. When the issuing of consecutive established laws intended to dampen the oscillations becomes too obvious a violation of the original plan (which promised to increase freedom and not the yoke) and can no longer be "justified," the issuing of secret decrees commences and secret actions are taken that violate the existing laws and constitution. The aim is always to preserve the official version of phenomena and to prohibit a revision of the forecasts that were wrong. No lawbreaking can be found in the officially permitted information. Therefore, more and more areas of social life are declared state secrets, which covers up the actions aimed at saving the integrity of the system at the cost of individuals' freedom or life. There is no trace of a "tyrannical whim" here; the process is logically continuous and consistent, because there are always only two alternatives: stop the oscillations by violence or change the structure of the system itself. Since the latter is out of the question, the use of force is inevitable.

This situation promotes the forming of different viewpoints in the group that is the focal point of all the feedback links in the system and makes decisions regarding all the parameters of economy and governance. The main fault line runs between those who think that the increasing use of force is unacceptable and that changes should be made in the existing structure and those who demand the use of force with no restriction. "Deviation" is obviously a relative term, but it will be used to label those who lose the power of decision-making, that is, who are removed from power. We must stress that it is not always the advocates of structural changes who are objectively right, since some changes could worsen instead of improve the situation.

It is also understood that improving the operation of a society, which is an immensely complex dynamic system, is always more difficult and requires greater intellectual effort and societal work than maintaining the status quo by using more and more force. But ossification of the oscillating system is highly detrimental to the cultural and technological development of a society. It is always with best intentions that rulers try to promote that development, which occurs through the creation of cultural, artistic, and scientific values. As a rule, great scientific discoveries, great pieces of art, and great technological revolutions are made by individuals or relatively small groups of experts. Societal practice, however, makes individual creativity difficult if not impossible, because it requires originality that in turn manifests in breaking with the existing customs and technological, scientific, or artistic conventions, which always threatens to perturb, deviate from the average, increase fluctuations, and, ultimately, promote oscillations in social dynamics. But since society fights against oscillations by any possible means, practically all originality is banned. It is all the more understandable as the plan is supposed to know the future, which leads rulers to attempt imposing on the stream of societal phenomena such a character that makes these phenomena foreseeable, since the unforeseen cannot be shaped and controlled. Hence the liquidation of all manifestations of individual activity which makes the forecasting more difficult. An optimal social structure should establish only a framework for initial stages of individual development and mutual dependence, the maintaining of which is necessary for harmonious coexistence of members of the community, but the universalistic construction plan of a centralized system tries not so much to give this development free hand as to shape it. Therefore, the plan implicitly excludes inventions, discoveries, works of art, and, in general, any values that it cannot foresee. This is precisely why cybernetics was a prohibited branch of research: the universalistic plan did not foresee the possibility of development of a scientific branch with such a broad range of influences in technology, biology, psychology, and sociology. The universalistic tendency of the construction plan on the one hand, and the practically inevitable necessity of using force to dampen unforeseeable oscillations in social dynamics on the other, bring about a society in which individual behavior is highly uniform; in which it pays to be average and unoriginal; and in which, to enable and support production processes devoid of automatisms, two enormous institutions grow beyond any limits: the pyramid of bureaucratic control and the pyramid of repressive apparatus. Societal labor thus feeds two machines of coercion. This is the reality of the plan for collaboration of the free with the free . . .

HYLAS So after a phase of relative dynamic equilibrium at the beginning, the system imperceptibly enters a phase of forced equilibrium, correct?

PHILONOUS Not exactly. Building a house requires many forces acting in certain directions, and they naturally cease once the house is finished; it would be a poor house if it needed additional support, i.e., continuous use of force to maintain stability after it is finished. A new societal system also requires force to be built, but when it is finished, no force should be needed to maintain its dynamic equilibrium and internal stability.

HYLAS Is this phenomenon limited only to the construction of societal systems?

PHILONOUS In principle, yes. When astronomical observations were inconsistent with Newton's theory (e.g., the shifting perihelion of Mercury), it was of course possible, instead of searching for a new theory that would explain the facts (as Einstein's theory did), to make the facts agree with Newton by falsifying the astronomical calculations, for example. But then the new theory of gravity would never have been born. In aeronautics, the first jet planes often crashed because of self-induced vibrations at high speeds. If the engineers had blamed the pilots instead of going back to the drawing board, the theory of induced vibrations in mechanical systems would not have developed and building airplanes free of those vibrations would never have become possible. It is extremely important that the failure of a first trial or experiment on the path toward the goal (the construction of a jet plane or of a new kind of society) not be taken as proof that the goal is principially unachievable.

HYLAS Do you think, Philonous, that it is possible to build a new society without using force?

PHILONOUS No, it is not.

HYLAS Then individual freedom must be limited during such an undertaking. But how are the possible abuse, injustice, and misery of whole nations to be avoided?

PHILONOUS Some minimum degree of agreement must exist for the realization of any collective endeavor, and it is precisely the task of sociology to determine that minimum and equip it with feedback links that will stop the spread of uniformity beyond what is necessary for the construction work. Human qualities like honesty, diligence, kindness, initiative, originality, and the wisdom to make good decisions are all functions of a societal system and thrive when they are rewarded (I do not mean only in the material sense). Postulating them with no regard to the objective laws of the system would be equivalent to telling soldiers under fire on a battleground to love their neighbor. It would sound like mockery.

Looking around in a capitalist system, you see everywhere signs of the feedback that we call making a profit, but in the centralist system you will find no universal manifestation of meeting people's needs. Yes, there are some attempts to realize this directive in the free education system, in universal health care, and in the respect for those who perform hard labor, but at the same time you will see fetishization of production, delayed and faulty supply-demand feedback, and the supremacy of institutional interests over the interests of individuals. So in some areas, feedback acts to satisfy people's needs; in others, a different type of feedback, unplanned, unintended by the theory yet as real as the other, limits, suppresses, and deforms people's needs instead of satisfying them.

HYLAS Do you see a solution to this problem? Is there hope?

PHILONOUS The initial conditions for the construction of a new society are few—the nationalization of the means of production, the abolishment of the private ownership of large agricultural lands, and the establishment of a system of general directives (a plan of development)—but in many areas a multitude of solutions are possible that differ greatly from one another, and we cannot say today which are better and which are worse, because the respective models have not been tested yet. The enormous responsibility of performing this kind of experiment is obvious as it involves organizing interpersonal relationships according to a selected structural plan, building a societal model that history has not yet tested, and running it for at least one or two decades, a time sufficient to observe the oscillations in the system to manifest as objective laws rather than as random and temporary deviations in some parameters. Only after this period will it be possible to answer the question posed by the experiment, "Yes, this is the right structure" or "No, it is not; we must reject it and build another with different parameters." It is clearly better to abandon the failed model instead of using force to stop its oscillations, which might have been fully justified to vanquish those who wanted to thwart the experiment itself in its initial stages but not in the later stages, when the model is already functioning, thereby manifesting its specific dynamic rules, and force becomes a tool for suppressing an opposition that no longer exists. The vices of bureaucrats, the resistance of alibists and conservationists, and the passivity of society should not be explained away by formulas like "a remnant of the past"; they must be examined as a consequence of the system's internal dynamics. The accusation of the "human nature deformed by capitalism" should be made only after an objective examination of a problem's sociological origins has failed to find other reasons. Since an experimental study of the laws of a currently hypothetical system in a gigantic electronic brain is far beyond our present ability, huge human, moral, and material costs are unavoidable.

HYLAS   Then you believe that a proper reconstruction of the socialist system will eventually lead to a structure that approximates our "ideal model"?

PHILONOUS   I think it possible, but I don't know how many attempts, failures, and experiments, how many years and how much effort still separates us from that goal. First, consider that a society that finds itself in a dire economic situation in the course of an experiment is interested not in the best construction path to the ideal model but in the shortest path to solving its current problem—two different things. So after a negative result of the experiment, it may be necessary to accept some kind of a compromise solution. Second, consider also that just as there exists a threshold of maximum complexity for a network, there must exist a limiting maximum complexity for a social system. The number of elements that compose a social system may therefore not be an irrelevant factor. An optimal structure for a small country may not be optimal for a large country, and vice versa. A centralist system may function with fewer perturbations in a small community than in a large society.

HYLAS So you believe there is such a rule?

PHILONOUS So far it is a hypothesis backed by little data. It does appear that the same centralist model can be maintained in countries of different sizes using different amounts of force to dampen oscillations. In a large country more force will obviously be needed than in a small one, but from that we should not conclude that a ruler of a large country is more likely to be a tyrant than a ruler of a small country. In fact, this is just a manifestation of a certain dynamic rule, which is as objective as the one that causes the elephant to have thicker legs than any other animal that is smaller. Since the parameters of production and social life vary in different countries, it would be nonsense to assume that all can adopt the same way of constructing a new social system.

The third issue to consider is that a system that guarantees maximum well-being to all its members still does not meet all the criteria that we listed above. Well-being is obviously desired, but an ideal system should go farther and guarantee increasing degrees of freedom for individuals while maintaining the adaptability of the system as a whole. A model whose production machinery (based on social automatisms) faultlessly provides people with everything they need should be the beginning, not the end, of the system's development and of our work. Because, as we said at the outset, social systems are nonlinear, that is, their dynamic rules change in time, and there can exist no system that will function in an unchanged structure for an arbitrarily long period. Every change in production tools will cause a shift in societal parameters, and a stable model may turn unstable in a new situation.

People therefore face not just two paths, one leading to socialism and the other to capitalism, but an enormous number of possible paths. Many go from quasi-socialist initial states back to a version of capitalism (not necessarily individualistic—it could be, e.g., state capitalism); many, starting from the same initial position but following different directives, go to internally unstable systems whose existence becomes impossible without the use of force. Those are systems with self-induced oscillations and coerced stability. It is a fundamental misunderstanding to think that the human misery in such a system can be improved by removing coercion in governing and in production. The removal of coercion alone will only increase the oscillations, which, unchecked, may quickly lead to a catastrophe—equivalent to opening the steam valve of an engine without an automatic regulator all the way. Periodic oscillations in the political realm of a centralized system are a manifestation of the described phenomenon where a decrease of force results in such an increase in internal perturbations that the government, fearing the loss of control over social processes, will quickly feel obliged to renew the force. This process in which many intellectuals see the demonism of rulers is actually a common dynamic law of a system without automatic internal regulation. Removing coercion gives positive results if and only if a system is at the same time undergoing structural changes aimed at the creation of links that automatically regulate social processes.

Finally, there are paths of social construction that lead to stable systems that guarantee increasing production, consumption, the satisfaction of all needs, and individual freedom, but may not guarantee further development toward the next forms, no less stable, of people's social coexistence.

HYLAS   This distinction puzzles me.

PHILONOUS   Those latter systems are systems that are highly sensitive to changes in the means of production. Here is an example. The complete automation of production, which principially rules out the persistence of the capitalist system, causes problems for the socialist system too, in that it imposes the necessity of creating new societal and individual activity goals beyond the production of goods for consumption. Automation therefore must advance hand in hand with a smooth transformation of a large variety of occupations and human activity in general toward increased individual and communal creativity that has no connection with the production of material goods. This contrasts with our world today. As you see, the ideal social system must possess not only social automatisms but also certain "stability reserves," stores of internal resistance to changes and perturbations. We are talking not of force but adaptive freedom and flexibility.

The present time, my friend, a time of the first trials and errors, is a moment in our history that is as tragic as it is heroic and will probably affect the future of our species more than any other period in history.

Reason, kindness, and courage—these are the qualities most needed now.

HYLAS   You are saying that humanity is now entering a time of trials and errors in the construction of a better world? It is hard to imagine millions of people transformed into guinea pigs for an experiment whose test tube is the whole globe. But why must there be costly mistakes if we can construct social models in a giant electronic brain, as you were saying? Shouldn't people wait until they have acquired the necessary theoretical and technical knowledge on theoretical models before conducting trials on the scale of nations?

PHILONOUS   An electronic brain representing a model of a social system would have to have a complexity equal to that of a system composed of elements each of which is a neuronal network; in other words, it would have to contain about a trillion elements. It will take more than a thousand years for us to construct such a thing. We must therefore resort to simplified models. What does this mean? That an electronic brain simulating a society, instead of operating with elements as complex as neuronal networks, would rely on certain "synthetic parameters."

HYLAS   I don't understand.

PHILONOUS   When we want to determine the path of a celestial body using the calculations of an electronic brain, we do not have to create inside the brain a model as complex as that body. The body consists of quadrillions, quintillions of atoms, but instead of providing a one-to-one representation of each atom, we can give the computer a series of synthetic parameters: the body's mass, its position and speed with respect to other bodies, and so on. The speeds of the individual atoms of the body will of course differ from our average, but in practice these differences are insignificant. We do the same with the "human atoms" in a society: since their one-to-one representation in an electronic brain is out of the question, we make an inventory of the parameters of social life and select only those that affect the dynamic rules of the collective essentially, representatively, and decisively. But there is danger in this. The brain does its calculations in the framework of a particular theory—for the celestial body, the theory is astronomical, namely, the theory of gravity. Had we used the Newtonian version, we would have found small deviations from reality, deviations that would disappear if we used a better theory: the Einsteinian took into account parameters that the Newtonian did not. Regarding society, the parameters essential for the dynamics of one system may be irrelevant or insufficient for predicting the dynamics of another. Certain personal mental characteristics or abilities may play no role in one system. If we omit them from our initial set of parameters, and with the aid of the electronic brain, we work out a seemingly "ideal" system, it may turn out that the system is not ideal in reality, because we selected the wrong parameters. Not to mention the nonlinear character of social systems, which significantly increases the difficulty of our task. Even when someday people can build an electronic brain that can serve as a model for sociologists, it will only be able to provide negative conclusions (i.e., that the given social system is not good), not positive ones: we will never know for sure that a model system that works fine will also work fine in reality. But even such simplified models are at least a century away, and humanity cannot wait that long.

HYLAS   So how do you see us proceeding?

PHILONOUS   Experimentation is necessary, as I said. But we must be cautious, just as with any other of our creations, whether an electronic brain or an atomic reactor. Traps lie ahead on the long and strenuous journey of societal experimentations. First, considering the enormous plasticity of human nature and its ability to adapt even to the most peculiar, difficult conditions of life, social structures do not reveal their objective drawbacks quickly and easily. And the border between objective and subjective is not always clear here. It is easy to blame the faults and imperfections of "human nature" for the failings of a system and then use force or other kinds of coercion to suppress this "nature." Instead of fitting the system to human characteristics and needs, we do just the opposite: we fit the traits and needs of the people to the system. This is the danger of the "bed of Procrustes."[8] When we fit people to the system, an accord between the theory and practice can only be achieved by coercion which spawns a lie. To various degrees people are aware of the lie: some see it clearly, some only suspect it, and some accept it passively, without even a thought about why it is happening. With time, the outcome of this mass pretending is that everything in human behavior that was coerced becomes an automatism, a convention. Pretense, like a mask that grows into the face and cannot be torn off, cannot be peeled away from the personality because it becomes an integral part of it. In this way serious deformations of the psyche arise.

HYLAS   What lie do you have in mind?

PHILONOUS   In principle, there is no social system, or at least there has not been one yet, whose operation does not include a lie as an element of collective processes. The lie hides the real motivation for a behavior, at the level of the individual or the collective, in the area of economics or politics, in relations domestic or international. But the participation of a lie in social processes is probably greatest in a centralist system, where all spontaneity of human responses is replaced by an organized response that the authorities impose. Community is required to response to every event in a unified way determined by theoretical tenets—and it really does that. Even those collective behaviors that in the old social system used to be expressions of true spontaneity (e.g., street manifestations) are now planned, staged, and obligatory. Public opinion, now a completely passive mirror of the actions and values of the authorities, becomes a complete fiction. An outsider who suddenly finds himself in such a society gets the impression that he is in a staged play and keeps waiting for the actors to finally remove their masks and stop acting—which never comes. . . . With time, such pseudo-spontaneous collective activities, which are just another consequence of the universal penetration of the authorities into every sphere of life, will form the image of an ideal citizen that, to real people, is like the wedding pictures with poses of wax mannequins in a provincial photographer's store window are to ordinary passers-by or like the norms in a *savoir vivre* handbook are to the ordinary human behavior. We can find at least three negative aspects in this. First, it narrows the individual options for response and squeezes personality into a corset of external directives concerning not only behavior but even emotions. It leads to a novel paradoxical situation when a citizen that actually responds in the way prescribed by the authorities, but spontaneously, is looked upon with suspicion, because the next time he acts in public, his response may not be in line: any authentic action prompted internally instead of guided externally is a potential danger to a system of collective pretense.

Social life, which under such conditions becomes a never-ending theater performance, favors those with acting talent, while those who do not participate feel that the human species has suffered a demonic possession, that personality has been destroyed in the name of goals that are incomprehensible. Yet this process is a simple consequence of what we have discussed: the divergence between the real laws of the system and the rules assumed a priori. Because the authorities dictate the responses that they think should follow from the tenets of the plan, every man lies to every other and ultimately to himself. He will not say that there is no bread; he will say that he is not hungry. Developmental possibilities, personal abilities, and basic longings are perverted. The bed of Procrustes means all of this but additionally also the use of coercion as a motivation to act not so much out of the desire for an award as for the fear of a punishment.

Second, it is a very difficult moment when one realizes that a modeled system has failed the test and must be rejected to make room for the next trial. After oscillations have caused a dire economic situation and great human misery, a collective need arises to find those who are responsible. In reality, of course, the culprit is the system itself (if we can call a dynamic structure a culprit) with its objective laws—but this is not an explanation that will satisfy the natural hunger for justice or even revenge. So certain groups are blamed—intellectuals, Negroes, Jews, party members, whoever happens to be at hand. This blaming can channel human passions—anger, disappointment, despair—to set back or destroy a social system, with the loss not only of the existing material values but also of the informational values of the experiment, which a rigorous sociological analysis could use for the future's benefit.

Finally, there is the danger, perhaps the greatest of all, of getting stuck in a blind alley. An example would be a system that limits individual freedom, thwarts personal development, and prevents talent from flourishing but at the same time demonstrates considerable internal continuity and stability and exhibits an overall increase in production, a resistance to internal oscillations, and a just and universal distribution of goods. Such a system may impose uniform thinking and behavior not through physical or psychological coercion but through the conventions it establishes in philosophy, the arts, pedagogy, communal and family life, and so on. Members of such a society, locked in the armor of these conventions since childhood, could lose all traits of individuality in a way reminiscent of the anthill.

HYLAS Is this not another bed of Procrustes?

PHILONOUS Not really, because there is no perceivable coercion in the system, so it is not easy to recognize the harm done to the individual. And if the members of this hypothetical society eventually manage to take it apart in order to build a new, better world for themselves, they will have to endure a long developmental delay, and the moral-psychological consequences will be so severe that it may be necessary for them to go through a transition phase, who knows how long, of reeducation in the values of freedom, courage, and initiative in thought and action.

Regardless of the selected path and the model being constructed, a considerable fraction of society members will probably realize the experimental and relative nature of individual and societal behavior since what is being built by the common effort is an analogue of a material or mental human construct and therefore a creation that is subject, like any scientific theory or a new machine, to analysis by inquisitive and doubting minds. But others will absolutize, fetishize, and mythologize their behavior by considering it faultless, which invariably takes people off the path of continuing development into a blind alley.

HYLAS I read somewhere about the possibility of constructing a "governing machine," an electronic network that would steer all social processes as a supreme authority. What do you think of this concept?

PHILONOUS It has been discussed by some cyberneticists but not as a practical possibility. The thing is that human coexistence in the capitalist system is not so much a collaboration as a competition; therefore, it manifests numerous elements of a game—John von Neumann studied behavior in this way. Rivalry or competition is a feature of a game where winning is wealth and success in life and losing is poverty and failure in life. Obviously, I am simplifying. A "governing machine" would be one of the players in the game, with the advantage that it would have access to information about all the "moves being played" and about all parameters of the game. The human players would not have that access. The machine could therefore predict the statistically most probable next state and, on the basis of that knowledge, make moves that would force the other players to submit, because otherwise they would lose (i.e., fail in life).

The governing machine thus acts on the principle of economic coercion like the capitalist system, which is no coincidence. According to some, it could save what the capitalist economists cannot, namely, the capitalist system itself. Others see in it a kind of "electronic Antichrist," a Moloch that would cause total social uniformity and create a state in which "statistical well-being" would be achieved at the cost of individual identity. Such a machine could be built someday, but one could just as well consider building a machine for psychological torture, so I do not see any sense in doing that. People need not an electronic governor but a better social system.

HYLAS   What is the difference between cybernetic sociology and traditional sociology?

PHILONOUS   Cybernetic sociology does not really exist yet; there are just the first buds of it in research. The science will arise only after the general theory of information develops its mathematical apparatus and accumulates a body of observational and experimental data large enough to make generalizations. Capitalist sociology and economics made many interesting discoveries in the past decades, but their fundamental limitation is that they treat (usually implicitly) the capitalist system as the only one possible. This is understandable, because capitalists, while often progressive in the area of technology, are conservative when it comes to systems, thus they like to wield established laws (of the type "Whosoever attempts to subvert the system will be punished"). No law prevents the reconstruction of production machines, but attempts to reconstruct the social machine are prohibited, therefore the experimental sociologist will face persecution, which is one of the reasons for the neglect and a lack of progress in this area.

HYLAS   And what does the cybernetic analysis of democracy look like?

PHILONOUS   Democracy means that feedback exists into which every citizen is "plugged" and so to some extent can influence the fate of the society. But there are two limitations to that influence. First is the steering of social processes in the existing system: what an individual can do (and what the entire society can do) is constrained by the society's objective laws. What in one system can be realized if the majority approves may, in another system be unrealizable, an empty resolution. In democracy, the realization of what the majority wants is possible only when material and structural conditions are in place. Further, people may vote for measures that are not to their benefit, for example, measures that lead to the rise of feedback that induces harmful economic oscillations. Therefore, a democracy, absent the appropriate theoretical knowledge of the social consequences of a change, may cause the most serious perturbations. The universal right to drive a car would result in a catastrophe if people were not required to learn how to drive first. So if people lack appropriate knowledge, a democracy can throw a country into a disaster just as a tyrant can. Introduction of feedback links that create conditions for democratic governing therefore is not a sufficient condition for the construction of an ideal model; specialized knowledge is also necessary. Such knowledge in a capitalist system is replaced by platforms of political parties, which are a poor substitute for a scientific theory of sociology.

Second, a true democracy should permit its citizens to approve also a complete change of the existing social system, which, since the foundation of capitalism is private property, "democratic" constitutions often disallow. As you can see, democracy as a formal method of governing is a function of the objective laws of the system and equally of its established rules. The established rules can coincide with real social processes, particularly in the area of ruling the country. Marxism found, a long time ago, that even the most democratic government cares mainly about the interests of a certain class, a certain fraction, of its citizens. If, for example, a percent distribution of participation in government is arbitrarily limited by law (such as the electoral laws), that is, "the power of decision-making" does not objectively reflect the present structure of personal relations, this distribution will shift, even against the letter of the law, through centralizing or decentralizing tendencies, until the decision-making reaches an objectively stable state. This phenomenon manifests differently in different systems—consider, for example, the influence of big capital on a government's decisions, which to a large extent makes "a government by the people" a fiction.

HYLAS   People form their social relations, through which, in a sense, they can change the existing system. But the system in turn formatively affects its people. Do you think, therefore, that human nature has any characteristics that are "permanent," "immanent," or "suprasystemic"?

PHILONOUS   Besides the objective laws, which derive from the social structure, the behavior of people as elements of this structure includes their individual psychology. The extent to which these individual features can be expressed is determined by the structure. Neanderthal society also had its Einsteins, but they were occupied with making fire, not with the theory of relativity. A social system can, either by the use of force or through the power of solidified cultural conventions (customs), prevent the full expression of personal abilities. As for the "permanent" characteristics of human nature, we need to distinguish what is inborn from what is acquired. The ability to talk is inborn; the language that a person uses is acquired and dependent on the environment. A social system should enable and facilitate the expression of individual abilities instead of putting people into rigid and predetermined molds imposed by conventions. Opening possibilities and increasing freedom—these should be the goals of social engineering. Scholastic debate about what is inherited and what is acquired, what is permanent and what is conditioned by environment, should be decided by practice. To decide empirically what depends on the structure and what on the individual we should first change people while keeping the structure constant and then change the structure while keeping the people identical.

HYLAS   Since human nature is on the table now, Philonous, I must share with you a thought that has bothered me for a long time. Don't you think that equating a human being with a network that can reason is an exaggeration, considering the cruel and reprehensible carelessness with which these networks were put together? Some have defects in their circuits for logical reasoning; some lack a feedback device for self-control so that instead of pursuing a goal they wobble mindlessly through the thicket of life; and some are missing the internal equilibrium that would enable them to form a coherent picture of the world.

The whole collection of these networks is swarming with individuals that are botched, uncoordinated, spastic, dense—if not outright stupid! Don't you think that the average physical differences among races and nations are nothing in comparison with the mental differences within the same race or nation, or even just the same family? By mental difference I mean not creativity but simply the ability to make use of the cultural products found in one's society. Does not the blind coupling between two human beings drawn together by the selfish pursuit of momentary pleasure and the mass production of the defective unfortunates jerked by life's contradictions betray all those high ideals that the human has dreamt up for himself and strives to pursue? Should not he who contemplates the development of his own species think less about the transformation of social systems and more about the biological reconstruction *generis humani*?[9] Not to multiply geniuses but just to make the mass production of morons impossible?!

PHILONOUS   What a fiery denunciation of humankind, Hylas! I did not expect such a thing from your mouth—after all we said.

HYLAS   You think I'm wrong?

PHILONOUS   Not at all. It is evident that stupidity exists in the world and that sociologists must make it a parameter in their equations. It is also highly probable that someday, in the distant future, people will set out to improve their biology to prevent inborn physical and mental disabilities. Yet your prosecutorial tone suggests not so much a philosopher's concern about the fate of our species as an intellectual's disdain for the plain bread eater, which, my friend, is a perilous and reprehensible attitude. This makes me think it will be a good idea for us to meet again in this park to conduct one more, our last, discourse.

VIII

PHILONOUS We discussed the construction of a society, tacitly making the simplifying assumption that all people who are the building material of such construction have essentially the same characteristics. Today we discard this simplification and examine what makes one human different from another. The subject will therefore be the structural quality of a social system, which more or less corresponds to the study by regular engineers of structural suitability and material strength. The motivation for our discussion is your last statement, which may have been a not entirely conscious expression of intellectual snobbery. You saw intelligence as a criterion for categorizing people, making that value judgment as if it were a fact. You went so far as to deny the designation of "human being" to one who does not reach a certain, not clearly defined, intellectual level.

HYLAS Isn't this criterion just and objective? Is it not obvious that a person should occupy a place in the social hierarchy, or an occupation, according to his abilities, intelligence, and talent? Don't you consider this to be an ideal for any social system? Is it not a fact that humanity owes its progress to people who distinguished themselves by their intellectual abilities and intelligence? Is not all human culture the product of intelligence, and therefore should not intelligence be considered the most valuable human quality? What in your opinion is a better criterion for categorizing people?

PHILONOUS First we need to answer the question whether criteria for categorizing people in the way that leads to a value hierarchy, according to which some people are more valuable and some less, are necessary, and, if they are, what goal they are supposed to serve. You spoke about such differences to justify the necessity of the biological transformation of our species. It is a bold and exciting idea, but surely it will not be realized in a future that can be foreseen with just a minimum level of rigorousness—because we lack the necessary knowledge, if not for anything else. Consequently, masses of people with average abilities will continue to live and share the same social system, for a very long time, with a great many people who are below average. This is a simple fact. But if we make these differences the guiding principle of our social engineering, it will only lead to the development of yet another theory of social elitism—this time based on intelligence—with the shameful result that intellectually disadvantaged people will have their rights curtailed, unless we consciously and actively prevent it from happening. And we can never be sure if some zealot of rationalism and a worshipper of intellect will not eventually come up with the radical idea of depersonalization camps, perhaps even extermination camps, for the so-called imbeciles with whom, according to you, the world is swarming.

HYLAS   I did not say or even think such an awful thing, Philonous, and you should not attribute to me such monstrosities.

PHILONOUS   I am not attributing to you any methods or thoughts, Hylas, merely taking what you said to its logical conclusion. You could be blamed for the mentioned extermination proposition just as the theorist and philosopher Nietzsche was for the practices of Nazism. The biological reconstruction of human beings is, for now, a fiction, yet your words, though meant only as a complaint, pointed to segregation of humankind, in which the feudal criteria of birth, the capitalist—of possessions, or the Nazi—of race, are replaced with the criteria of intellectual ability, but it can only end up valuing some people over others. I vehemently oppose this. Any construction of a social system that enables the establishment of a privileged elite must have nothing to do with our aspirations and explorations.

HYLAS   Your opposition has little justification if it only stems from causes that are emotional and moral in nature. Moreover, it contradicts the objective fact that intellectual differences do exist.

PHILONOUS   Of course, my opposition is on moral grounds. As we said, social engineering must be subject to ethical criteria, because here the building blocks of a system are people. But what I say next is a mix of ideas and facts based on science.

HYLAS Go ahead. I am listening.

PHILONOUS We must first settle, as rigorously as possible, the matter of human intelligence. For ages people have been debating whether intelligence is an inherited or acquired trait. Research, particularly from the last decades, suggests that intelligence, that is, the maximal intellectual effectiveness that enables a person to handle all kinds of professional and life situations (obviously, this is not a rigorous definition!), is determined by both genetic and environmental (epigenetic) factors, the genetic having greater weight (although some utopianist socialists don't want to hear that). At least in principle, the differentiating effect of the environmental factor can be eliminated so that we obtain a state determined purely by heredity. In a society with a significant degree of structural inequality, in which the privileged classes hinder the development of those from the lower classes, the average intelligence in the lower classes will always trail behind that of the ruling classes, but it is for environmental, not hereditary, reasons. The further the progress of democratization, that is, the process of equalizing the starting position in life for all members of the society, the smaller that difference will be. A limit to this process is a state (which has not yet been achieved anywhere on earth), in which environmental factors will have practically no effect on the statistical distribution of intelligence in the population.

HYLAS No effect at all?

PHILONOUS No, I said they would not affect the statistical distribution of intelligence in society. In other words, they will cease to act as differentiating factors. When an environmental factor influences everyone equally, it stops making a difference. Where a selection based on belonging to a caste or a race does not hinder access to education, information, and ideas, the environmental factor can be removed from our calculations. It will still be operating, of course, but equally on everyone, and therefore the difference between people in intelligence will be due only to heredity. In this ideal situation the distribution of intelligence would be represented by Gauss's normal, bell-shaped curve: the largest number of individuals have average intelligence, and the number of those with higher or lower intelligence decreases as the distance from the average increases. It follows that intellectual inequality, having its source in nonenvironmental, inborn, or inherited factors, is a fact that the engineer of social systems must take into account. Rejecting this fact, even out of the noblest motives of humanism and egalitarianism, will manifest sooner or later and lead to significant disturbances in social practice.

HYLAS So, *nolens volens*, you end up agreeing with me?

PHILONOUS A little patience, my friend. Let us consider where intelligence research and measurements came from, what they are like, and what purposes they serve. They originated in a field called psychometrics. After years of trial and error, experts processed an enormous amount of data from what is commonly understood as tests designed to differentiate people according to certain measurable traits of intellectual ability. It may sound paradoxical, but the usefulness of these tests, that is, their social value, is unquestionable, even though we do not know yet what exactly it is that they measure—there is no agreement among the experts on this. Yet the differences that the tests uncover significantly correlate with practical experience in life. For example, predictions of success or failure in the school performance of young people based on intelligence tests have been accurate throughout the existence of this science. True, the tests do not explain the psychological mechanisms behind intelligence; but the same applies to a thermometer, which says nothing about molecular processes yet correctly measures temperature, regardless of which particular theory of thermal motion the user of the device favors. The tests, exactly like the thermometer, are a measuring tool. There are many kinds of such tools, some of which measure the so-called overall intelligence and some of which test for specific sets of abilities that point to a particular job for a particular individual (e.g., mechanical, clerical, or mathematical skills). This field of study has developed mainly in the United States—in the form of highly specialized branches of industrial and social psychology, professional counseling, and so on. That these tests are now used all over the world in disciplines that could hardly do without them proves their practical value. In aviation, for example, pilot training is so costly, the imperative of having an air force is so powerful, and the necessity of the most rigorous selection is so evident, that tests are used to secure the optimal quality of human resources in this field.

The tests have a high diagnostic value, able to determine quickly and objectively the distribution curve of intelligence in any given society, as well as the mental power of its members (with better accuracy and precision than accounts of lifelong acquaintances). They also have a high predictive value for an individual's success in a field of study or in a job. The work with children of six has less predictive value than the work with adolescents of sixteen, for whom the predictive value approaches 90 percent, which shows that personal intelligence changes little later in life.

Everything said so far assumes that the administered tests reliably measure certain traits of mental ability and that the subjects are equally familiar with the tests, because a person acquires proficiency in taking tests, and so taking many would affect the score. Of course, there are tests that have little or no value, in which case raw experimental data and their professional evaluation by experts become more important. There are also tests, especially for character and personality traits (e.g., Rorschach) that are not as objective as the intelligence or skill tests, where the results depend to a significant degree on the discernment, insight, and intuition of the examiner, which is obviously a severe drawback.

Let me make a few remarks about the overly bold generalizations and groundless extrapolations based on the psychometric methodology and results often used by psychologists. A methodological analysis of these claims shows that they amount to the stratification of human society according to characteristics that persist in its subsets. In every larger ensemble we find the normal, Gaussian distribution of intelligence. There will always be people with intelligence well below the average and people with intelligence well above the average. The average of the ensemble may not equal the average of the whole society—at a university or scientific institution, for example, the average will be higher than in the entire society. What interests me here are the individuals with the highest intelligence. Psychometricians tend to define a genius not as one who appears rarely and who possesses extraordinary creative ability in a particular discipline but as one whose intelligence quotient is as far above the average as possible. This is a serious methodological error. First of all, there are no "absolute" units of intelligence; a numerical value accepted by convention is the result of mass studies and is therefore related to the statistical average of the data. The application of IQ loses scientific meaning when it is made beyond the statistically probable limits of the intelligence distribution, in other words outside the region of valid experimental data—it is like looking for the Earth's poles on a map in Mercator projection.[1] Second, the tests do not measure creative ability in principle. Attempts at developing a test for creativity have not gone beyond the stage of initial, preliminary, and not too successful experiments. Psychometrics is just a particular kind of measurement, and any measuring tool can be used only within a certain range of parameters. A thermometer that correctly gives the temperature of water will lead us into error if we use it to measure the temperature of a cosmic gas. In the same way, a test designed to examine a statistically common trait will fail if we use it to measure a trait that is practically nonexistent in a society, such as an extraordinarily creative ability.

HYLAS   I see the value of tests as a tool for selecting an occupation, but to make the course of an entire human life—by pushing a person into a particular occupational path or, conversely, blocking that path—depend only on answers to a dozen questions given in an examination hall, on a preprinted form in pencil, hardly agrees with the tenet of an individualistic and humanistic approach to people's problems. It can happen, and surely often does, that a test will indicate an intelligence lower than what the subject really possesses because the subject was stressed or "blocked" during the examination.

PHILONOUS   You echo reservations that are now almost classic. It is certainly true that a person's intellectual abilities and emotional life cannot be separated. If a subject, taking a test to become a high-voltage technician, gets a low score for emotional reasons, we might expect that he will not respond well in a real emergency, such as a power line accident, when quick decisions and actions are required. A test would be inadequate exactly if it failed to reveal such weakness in the subject. In selection, the important question is whether or not a candidate is suitable for a job; the question of why the candidate is unsuitable—whether the reason be emotional or intellectual—is unimportant but can be answered with additional specialized tests, if needed. As for the inhumanity of deciding a person's fate on the basis of a few hours of testing, I would just point out that the predictive value of psychometric tests is on average four to ten times higher than that of any nonobjective method, though the nonobjective method may be kinder (a letter of reference, the opinion of a friend or acquaintance, etc.). Opposing the results of experiments that count in hundreds of thousands is groundless and foolish.

After this praise of psychometric methods, let me now criticize their use, especially in the United States. My main criticism is that such testing is used for selection in an open system.

HYLAS   What do you mean?

PHILONOUS   An example of an open system is a biological population. Evolution takes place in it by selecting out less adapted individuals, who are condemned to extinction and will disappear after one or more generations. The system is "open" to extinction, rejection, or removal of the individuals identified as "inferior," that is, less adapted. A society is open when certain occupations and socially valuable positions in its structure are "protected" by a "test filter" that allows the entry of only individuals who pass the test; those who cannot are rejected like social chaff. In such an open system, tests aid in the formation of an "elite of the capable." But society should be a closed system, in which the purpose of testing is not to reject or filter out people but rather to determine their abilities. The most important function of a test should be diagnostic, directing people to a particular path of study, an occupation, a career, not filtering out those who are less capable. Systematic career counseling, long-term life path advising, and looking for the maximum of abilities—these are the tasks of psychometrics in a society that is a closed system, where no one is discarded or condemned to permanent unemployment. Only in this framework can testing increase the effectiveness of social processes, optimize an individual's potential, and thereby decrease the number of personal defeats and life failures.

In the United States, tests are employed according to the open system model: universities, corporations, banks, and offices use them as filters to fish out from the social masses the most capable and most productive individuals; providing professional counseling to the rejected candidates is not in the business's interest. As you see, it is not easy to use testing in a genuinely humane way, with the motivation of caring about the fate of the individual, in a society where the interests of the owners of the means of production collide with those of humanity as a whole.

But you should not think that selection takes place only where testing is widespread. The discriminating filtering of people for occupations and positions occurs in every social system. Often it is not under anyone's conscious control; it happens spontaneously when certain fields in a social hierarchy somehow "preferentially attract" people with certain traits and abilities. There is interplay here between the "requirements" of unoccupied positions in a society and the broad range of traits that are available in the population. Certain individual abilities, say, mathematical talent, could not manifest themselves in a Neanderthal society for the simple reason that there was no "social demand" for them. Conversely, a society may have a need for mathematicians or physicists—for example, in connection with a push to automate production—but because of the lack of people with required skills few of those vacancies will be filled. Yet this is trivial. More important is the fact that selection criteria can be not merit-based but instead established by tradition, hence obsolete and having nothing to do with the real, substantive requirements of the vacancies, or they may be dictated by an imposed political or religious dogma.

HYLAS What exactly do you mean by merit-based selection criteria?

PHILONOUS First, even the most objective method, that of psychometrics, is just a tool, and, therefore, as with any tool, it can be used to society's benefit or harm. For a drastic example of harm, a tyrant uses testing to select the most intelligent citizens in order to imprison or kill them, thinking, not without reason, that these are the people who pose the greatest threat to his regime. Second, in practice selection means privileging certain traits for certain positions or occupations. These can be traits that are truly necessary in a job—imagination and construction skills are necessary for an engineer or architect, mechanical talent and good reflexes for a driver or pilot, specialized knowledge and organizational skills for an agronomist in charge of large tracts of farmland—but there are also less merit-based traits, and these apply to all kinds of managers, for example, the ability to impose one's will on subordinates, regardless of whether the objective has professional justification or not. Being an opportunist, having a good memory for quotations, eloquence, and knowing how to flatter a superior are other examples.

HYLAS So you are calling traits not based on merit those that serve the interests not of society but of an elite or a ruling group.

PHILONOUS Basically. They can also, if not directly serving an elite, result from a doctrine that the elite is proclaiming. A selection filter of "racial purity" would be an example. A routine application of criteria not based on merit is found especially in centralist systems. Commonly called "cronyism," it fills all socially valuable positions and occupations with incompetent people, which leads to severe disturbances in the economics and dynamics of the entire system. But note that some resistance to selection criteria not based on merit, resulting from objective developmental factors, is practically unavoidable. If in any field of human activity a change takes place that demands new methods and therefore new skills, the people in this field typically defend the methods and views of the past by all possible means. Take psychology as an example. At the beginning of the twentieth century, mathematical skills did not belong among the traits required for the profession of a psychologist. The spreading mathematization of psychology, related to the revolution in the field sparked by psychometric studies, resulted in a significant change in selection criteria for skills of a psychologist. This makes us understand the distaste for psychometric methods and the mathematization of psychology in general among the psychologists who were trained before the change.

Let me mention one more consequence of discriminating trait selection in society that can determine the life path of an individual. Every system, whether it employs scientific, objective selection methods or the selection occurs spontaneously, contains some people who for various reasons are unable to meet the requirements for any of the available professions or societal functions. I am talking about "misfits," who can never adapt and are always filtered out. I believe that studying this group will allow us to draw conclusions about the overall value of a given social system, about the selection criteria that it employs, about the conflicts and tensions that typically occur in it, and so on.

HYLAS May I assume that this group usually consists of neurotics and neurasthenics?[2]

PHILONOUS The composition of this group will be different in different social systems. Misfits are always filtered out, but societies do not all use the same filter. In some, the maladapted may be individuals of high moral quality—for example, in a social system that resembles a Nazi concentration camp. These would be the people who cannot accept social roles imposed by force, who refuse, say, to serve as an executioner of their fellow inmates, or to work in a crematorium or at a killing ground. I am mentioning this just to show you that selection does not have to be based on intellectual differences. Intelligence is just one dimension, and selection in a social system is, as a rule, multidimensional.

HYLAS I understand your point about the multidimensionality of a selection process, but in a normal society the people filtered out are mostly weaklings, in both intellect and character—so neurotics would be predominant there, don't you agree?

PHILONOUS What do you mean by "normal" society? Is the Nazi system abnormal? In it, the filtered out may be people whom we would consider morally wholesome, yet the selection filter imposed by rulers favors the traits of ruthlessness, blind obedience, cruelty, and intolerance—to the extent that some experts argue that the top positions in such a system are held by sadists and psychopaths. But let us be careful here: reducing a sociological problem to that of individual psychology disregards the influences of a social structure. People can turn into sadists out of necessity, if a forcibly imposed convention demands it, not only out of a personal tendency. The thesis of Kretschmer, that normal people are ascendant over psychopaths in times of peace while in times of revolution and upheaval the psychopaths, becoming dominant in a society in turmoil, rule over the normal people, is undoubtedly false.[3] The cause of sociopolitically motivated crime is most often found in the very structure of the existing system, in its objective dynamic rules, whether they result from interclass struggle or the tyrannical behavior of the rulers in a centralist system. Explaining such crime solely by psychopathological analysis is a fundamental methodological error and absolute misinterpretation that harms the progress of science in general and sociology in particular.

As for neurotics, they usually are not born that way but become neurotics mainly as a result of their environment. This issue has a curious social aspect. In general, the prevalence of neuroses increases as a society improves. Indeed, stress and anxiety disorders seem to be a byproduct of a high level of civilizational development. Experts confirm this. It is known that not only neuroses but also severe phobias disappeared, that is, were "cured," in the concentration camps: anxiety disorders and obsessions subsided in the face of very real and terrible danger. Not that I am recommending this kind of therapy.

HYLAS I have always been puzzled why psychoanalysis, so popular in the United States, is not equally popular in other countries of the West, for example, France or Italy. Can the standard of living play a role here, since, as we know, it is the highest in the United States?

PHILONOUS That is likely one of the reasons but definitely not the only one. The problem of mental illness is just another aspect of the complex dynamics of social systems. It is fashionable for the affluent in North America to have "issues" with their subconscious and a psychoanalyst to deal with them. But this trend is more than a form of snobbery. There is no doubt that mental life, along with its realm of subconscious, can be cast in many different molds, and its manifestations can be interpreted in many different ways. When such conventions spread through a given environment, as in the United States, a majority of well-to-do people will eventually turn into a willing and welcome material for psychoanalysts. Such people produce dreams in accordance with Freudian theory and exhibit textbook "complexes," thereby verifying the tenets of psychoanalysis (including those that are doubtful) as the pansexualism of the subconscious, the prevalence of the Oedipus complex, and the fear of castration. Attempts to find a comparable epidemic spread in subconscious mental life in a society with different conventions, say, in a centralist system, would be fruitless. Thus the phenomena proclaimed by psychoanalytic theory constitute a closed circuit with positive feedback: patients fuel the psychoanalysts' faith in the validity of their theses, and the psychoanalysts in turn amplify the beliefs and symptoms in their patients. But it is not true that everyone's subconscious is populated by sexual symbols to be uncovered by a psychoanalyst; in reality, these phenomena are relatively rare and found only in a precisely defined environment (people who are well-to-do and fairly intelligent, or at least who believe they are such). Yet they can impose themselves on and spread through larger segments of a population, which then creates the impression of their universal prevalence.

In reality this is just a part of a much broader phenomenon—the imposition of specific conventions on society. The psychoanalytic convention is better than the medieval convention of witch-hunting, but only because it harms no one except the patient. In general, if one undertakes to impose a certain norm, convention, or way of life on a given society and does it with sufficient resolve and ruthlessness, the desired effect will be always achieved: the stratification of a previously homogeneous group, say, or the emergence of a new antagonism to cover up an antagonism more fundamental in that it comes not from propaganda, an irrational taboo, or an obsolete doctrine but from an objective perturbation in the dynamics of the social structure. The well-known ancient method of *divide et impera* is precisely such a use of imposition.

Bearing in mind the example of the horrible sociological experiments that were the Nazi concentration camps, we can conclude that through the use of coercion with no limits one can construct almost any system of interpersonal relations and impose on it any preselected stratification or segregation into privileged and disadvantaged sections or into a hierarchically ordered castes, where the only real benefit of being an "elite" may be a short delay in death at an executioner's hands. I am referring, of course, to what took place in the ghettoes during the Occupation. That phenomenon no doubt deserves careful sociological analysis.[4]

HYLAS I have to admit that these new aspects of the problems of coexistence in human society (particularly regarding conventions that can steer people's minds and selection criteria in a given system) have muddled the picture that I formed during our conversation up to this point. I mean the model of society as seen by cybernetics. I don't see how, or if, we can include into it the things that we are discussing.

PHILONOUS   I am not claiming that cybernetics is methodologically omnipotent because I have no wish to replace one generalization with another. Yet where is the difficulty here? All the phenomena that we have discussed can be interpreted as, on the one hand, instances of mental processes in certain neural networks and as, on the other hand, the transmission of information in the vast network that is a society. An adequately developed mathematical apparatus of information theory will be rich enough to model all this.

Let me summarize what we have established in our discussion of structures of social systems—from the perspective of cybernetic sociology, which is yet to emerge. Any activity that is taking place in a social system under coercion or repression has basically a single aim, to turn a nonlinear system into a linear one by the simplest means, namely, by decreasing the number of degrees of freedom available to the individuals who are the system's elements. Imposing on these elements prescribed norms of behavior (which should be as uniform as possible), and thus erasing the diversity in all individual responses, makes it easier to predict, regulate, and shape the future course of social processes.

Conversely, the more freedom the individuals are given, the greater the nonlinear perturbations in social processes, because the range of possible actions will broaden, contradictory opinions will appear, and there will be debates and opposing actions. As the system moves further and further from linearity, maintaining its internal cohesion and predicting its future will become more and more difficult.

Although it is easier to maintain order and keep all the processes of a system on a smooth course when force is used, and although people can adapt to living according to the norms and conventions imposed on them, data obtained from history as well as from studies and experiments in the last few decades clearly show that the mechanical effectiveness of a system and its linear character alone do not guarantee that its elements, people, will be happy.

HYLAS   So we must seek a golden mean between linear and nonlinear, between the maximum freedom of anarchy, which dismantles any structure in society, and the other extreme, the shackles of uniformity?

PHILONOUS   In a sense. I am just rephrasing what we already said about the necessity of bringing into one system the highest possible degree of individual freedom and the highest possible degree of internal cohesion based on the mutual complementarity of human traits.

HYLAS   I confess that I have grown pessimistic about the prospects of creating a mathematized sociology, understood as a general dynamic theory of social systems, because I can see all the trouble piling up on that path. You said yourself that a person's behavior, or, if you prefer, future operations and states of his neuronal network are not completely determined by its past states, and therefore predicting future states of a social system, no matter how deeply rooted the prediction is in a theory that is most mathematical of all, will only amount to listing a series of possibilities with varying degrees of probability. Talking about a "design" or "structure" when conditions are so fluid and developmental options so numerous—is that not groundless cognitive optimism? If we subjected a system of tyranny to a rigorous analysis, our cybernetic model would have to incorporate not only the fact that some information was blocked while misinformation was being spread in the society's information channels, but it would also have to factor in such psychological responses to tyranny as individuals' fear or cunning and—what would be the most difficult—it should contain the possibility of the emergence of an exceptional individual (say, a Marx) who would organize a theory-based freedom movement. What mathematical model can account for such a possibility? How to calculate the probability of its realization? I just cannot envision it.

PHILONOUS   These are not insurmountable difficulties. Let me give you a simple example of a rigorous, mathematical treatment of a sociological problem. Take three possible aggregate phases of social "atoms," that is, of people: tyranny, democracy, and anarchy. The imprecision of these terms, at the present state of our knowledge, is unavoidable. The x-axis indicates the number of the degrees of freedom possessed by individuals. Zero degrees of freedom would mean that a person is incapable of any action, that is, he is dead. Someone whose behavior is not constrained by the existence of others and who does not take into account anyone besides himself has an infinite number of degrees of freedom—Robinson Crusoe might be an example of that. We should, of course, distinguish degrees of freedom with respect to other people and those with respect to physical environment (including a civilizational level), but we will neglect this here for simplicity. The second scale under the *x*-axis is a measure of the force acting on the system: at the maximum, the smallest transgression of the existing norms and laws is punishable by death; at the minimum, there is no punishment for anything. Evidently, the stronger the acting force, the fewer the degrees of freedom. On the *y*-axis we have the "mental temperature," which is the tendency to act spontaneously, regardless of or even despite the existing laws, norms, and conventions (either imposed or voluntarily accepted) in the system. The second scale on the *y*-axis shows the amount of information that activates spontaneous action at the given "mental temperature" in the community.

The graph shows the three partially overlapping regions of tyranny, anarchy, and democracy. The lines that define the phases represent phase boundaries; crossing them signifies a transition from one phase to another.

Where the phases overlap, two or even all three systems can coexist at the given values of the parameters, but of course only as alternatives in any given society at the given time. In the single-shaded regions there is absolute sociodynamic stability of the phase. The overlapping

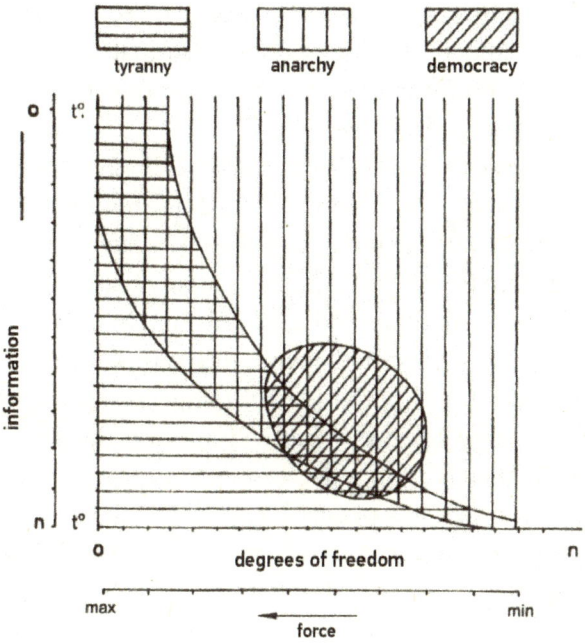

areas represent parameter values at which each of the phases has relative stability, with a specific (statistically determined) probability of transition to another phase. The graph, though simple and primitive, provides insight. Observe first that with few degrees of freedom, corresponding to a significant external force, the only stable phase at a low "mental temperature" is tyranny. As the number of degrees of freedom increases while the "mental temperature" stays low, we reach a critical point in tyranny's relative stability; the probability of a revolution increases, until, above a particular value of the number of degrees of freedom, it becomes a certainty—but the phase that replaces it can only be anarchy. If the degrees of freedom increase at a higher "mental temperature," we reach the region in which the transition may be to anarchy but also to democracy. Whereas both tyranny and anarchy have regions of absolute stability, democracy does not, which means that there are no parameter values at which the transition of either anarchy or tyranny to democracy would be a *certainty* from the point of view of sociodynamic probability. Above the critical value of "mental temperature," anarchy becomes the only possibility, because such a community will not be subject to any conventions, whether democratic or autocratic, that would organize the group of human atoms into a structure (governed or self-governed). Finally, there can be such a high number of degrees of freedom that only anarchy can exist, this time simply because there are *no* laws or conventions to regulate the behavior of individuals.

What else can we deduce from our graph? First, it implies that as long as a tyrant keeps the communal "mental temperature" below the critical point and at the same time does not allow the number of degrees of freedom to increase (i.e., the tyrant will use whatever force is needed), the tyranny phase can exist indefinitely, at least in theory. Second, democracy is a sociodynamic phase that is only relatively stable: an increase in either the mental temperature or the degrees of freedom above an optimal value will increase the probability of its transition to anarchy, and a decrease in these factors will lead to tyranny. Third, at a very low or very high mental temperature tyranny cannot transition to democracy; if there is any transition, it will be directly to anarchy.

Of course, all this is well known from history, both ancient and modern. What is novel is the use of an analogy from physical chemistry, more specifically from the theory of thermodynamic equilibrium between phases. Just as water can exist in the liquid phase above its boiling point only under elevated pressure, tyranny can exist above a particular mental temperature only under the elevated "pressure" of repression. And, just as there exists a value of high temperature (the critical point) above which no pressure can liquefy water, if the mental temperature becomes very high (e.g., when the oppressed feel they have nothing more to lose), even the use of maximum force cannot prevent the transition from tyranny to anarchy.

The graph even answers your question about the possibility of predicting the emergence of a Marx in a given society: the possibility is equivalent to that of the emergence of an unusually large amount of "triggering information" from spontaneous nonconformist activities. At a low mental temperature, the amount of information required to trigger a community may unfortunately be close to infinite. The graph indicates that change is possible only in regions whose stability or equilibrium is not absolute. In addition to the sociodynamic probability of a change (not depicted), we need to consider "socio-energetics." The overthrow of a social structure requires human effort. At the same temperature and freedom, it is generally easier to overthrow democracy and install tyranny than the other way around. The installation of anarchy requires the least effort. The transition from anarchy to democracy requires energy, but the transition from tyranny to anarchy may occur spontaneously (similarly from democracy to anarchy), because in that case all structure is being liquidated—an example: when a tyrant dies and his henchmen flee, chaos results.

HYLAS How can there be anarchy with a low number of degrees of freedom?

PHILONOUS That happens only at very high temperatures, where it is not the power of the government (there is no government) that limits an individual's action but the impulsiveness and expansiveness of other individuals (which will be enormous at such a high mental temperature). In terms of sociological consequences, do not take the upper part of the graph too seriously. Without any participation of social structure, mental temperature values can develop a pressure that lowers the number of degrees of freedom—for example, during a volcano eruption, on a sinking passenger ship, or in a city engulfed in urban warfare. In normal circumstances a community does not reach such temperatures, although individuals can. But let us return from our excursion to sociology's past into the present.

HYLAS   One more question, Philonous. It seems to me that the graph should also contain empty, nonshaded areas, in which no phase of social aggregation is possible. If the universal "boiling" of minds rules out a large number of degrees of freedom, as you just said, then a state that cannot occur should be called "forbidden"—by analogy to similar physical states (in quantum mechanics).

PHILONOUS   I see your point. States that you call "forbidden" would be those with an extremely low sociodynamic probability of occurring. This probability fundamentally depends on the community's history, which causes the nonlinear functional dependence of both our parameters in the first place. A sudden increase of freedom in tyranny (after the government's retreat and a release from repression) usually results in an increase in mental temperature and demands for more freedom; the government then either gives in or returns to repression. So we have here a rather complicated dependence with cyclical feedback loops (an increase in repression activates negative feedback, whereas decreased repression activates positive feedback, that is, enhances community demands). Yet sometimes this process may unfold differently, and the government's retreat fails to increase the mental temperature. What is then the functional dependence between the two parameters? Think of an animal in a cage. When we open the cage and free the animal, we grant it a greater number of degrees of freedom. But sometimes the animal, released, will continue behaving as if it were still in the cage, pacing just a few steps back and forth, because its lifelong behavioral conditioning is stronger than the new situation. A society that has been living under tyranny for a long time, especially a ruthless tyranny, may not exhibit a rise in mental temperature after the government retreats and prohibitions are removed. The people may not behave in a wild and nonconformist way. But as the government continues to weaken and a social cohesion that it imposed drops, for any reason, almost to zero, the society, being at the border of the tyranny's instability, can "explode" and transition to total anarchy in a single jump in mental temperature. Another society, having a history free of repression, will react in a completely different way to both the repression and its weakening. Hence the nonlinearity of a social system stems from the system's history and experience. Without taking history into account, our graph has no prognostic value, and I never claimed that it had any. To calculate the sociodynamic probability of a transition requires taking into account additional parameters, which is precisely the task for the future theory. Even the terms like "tyranny" and "anarchy" as signifiers of phases of social aggregation are just simplifying generalizations of many structural possibilities. Furthermore, in addition to mass-statistical aspects, the theory will have to take into account also individual aspects of these phenomena. As you know, a speck of dust dropped into water will not affect its state, but if the water is supercooled below the freezing point, the speck, acting as a crystallization nucleus, will cause a sudden change of water into ice. Similarly, an individual's role in the shaping of social processes or the reorganization of an existing structure depends on that structure. When a tyranny is absolutely stable, even the most nonconformist element will accomplish nothing except its personal liquidation. But if the stability is only relative and the number of degrees of freedom is high enough, the same element can precipitate an uprising. This too is a simplification because we neglected a society's class stratification, which also plays a role here. But enough of divagations, let us return to our topic.

HYLAS Why, when we seem to be on the verge of making some interesting conclusions? The role of an individual in social transitions surely deserves attention. Continuing your physicochemical analogy, I would call your approach catalytic. You say that an individual can significantly affect the course of history when the "potential for nonconformist activity" exists and the present state of the given community allows for it. The push he gives then to the social processes is like the bit of a catalyst that makes possible a chemical reaction. The given individual possesses special abilities, those of a catalyst. But imagine a situation when any individual, without any special skills, can affect the course of social transformations, namely, when he occupies a privileged position in the social structure. A king or tyrant can be a completely mediocre person but because all feedbacks in social processes converge in him, he has the power to make decisions on war or peace, and other important issues. Therefore, the "catalytic" effect of an individual on a society can take place only when the state of the system and the qualities of the individual are right, though "topology," that is, the position that the individual occupies in the structure, is also relevant.

PHILONOUS   Indeed, an individual can affect the fate of a society like this only in certain systems. An ideal system would make the "topological" and "catalytic" factors, as you call them, independent of each other. A society's fate cannot, and should not, depend on whether its leaders are "providential" or mediocre. Social engineering would not be possible if every construction depended on the chance appearance of exceptional people. The maximum degree of freedom in such an "ideal" system must not include the freedom of putting society's fate into the hands of one person. The maximum values of parameters (e.g., number of degrees of freedom) in a social system do not equal to the optimum values that ensure the stability of its structure and processes. Hylas, we could talk about this for hours. Let us instead return to our topic.

HYLAS   I won't allow that. Your refusal to admit the influence of exceptional individuals in a society is no different from prohibiting any deviation from linearity, and therefore it amounts to limiting the degrees of freedom in order to preserve the system's structure. How is this better, I ask, than what a tyrant does?

PHILONOUS   But, Hylas, if we construct an optimal system, we must protect it from disintegration or turning into a system that is worse (e.g., one that oscillates). The whole point is to embed this protection into specific automatic social processes instead of using force! Once a system's construction begins to be based on a scientific theory, there is no longer any place for human whim or multiparty politicking. Imagine the construction of an atomic reactor that is not carried out according to a unified plan based on the theory of physics but instead depends on competing "programs" thought up ad hoc by individual physicists. It will end either in a disaster or in nothing. Similarly, once we have determined that the best way to prevent polio is by vaccination, we must prevent charlatans from replacing the vaccine with "healing beads" and other nostrums. From the viewpoint of scientific sociology, politicians are amateur healers of social ills, practitioners who, lacking theoretical knowledge, are guided by intuition—at best. If, having recognized the role that science plays in all areas of life, we prohibit it from articulating rules that would be binding for everyone in the realm of social relations, then we will regress to the time before Marx, my friend! When an astronomer says that the Sun is 150 million kilometers from the Earth, is he "imposing" this on people? Your objections are methodological nonsense and serve only to show the enormity of the irrational resistance that needs to be overcome in people so that they will accept the objective sociological laws and everything that follows from them.

HYLAS   You argue that no statistical fluctuation in activity caused by an individual should be allowed to make a dynamically stable structure unstable. But as an astronomical theory must hold for as long a period as possible, so must a sociological theory. If we look into the abyss of future time that gapes before humanity, we must conclude that the current bases of both individual and collective action, rooted in the economy, will atrophy and disappear someday. Automated manufacturing, atomic energy, and their synthesis will provide people with a practically unlimited amount of goods. What then? New forms and contents of human activity, which will go beyond the economic—because when all needs are satisfied, economic stimuli are unnecessary—will unavoidably change social structure. And since no theory can foresee the psychological contents and the cultural and civilizational conventions that will exist after we become free of the economic motivation of behavior, we must accept that neither can the theory of social engineering do that. According to this theory, our "ideal" system will be only one of many *stages* on the path that leads to a society whose tasks and goals we cannot even imagine.

PHILONOUS   You are putting in different words what I have already said: no system of interpersonal relations can be eternal. An astronomical theory holds for billions of years; a sociological one, for perhaps a few centuries—precisely because society is a nonlinear system. Yet, although we cannot predict all social processes and states, our theory is not as weak as you think: it can entertain the *possibilities*. Although we do not know what collective human activities will be like ten thousand years from now, we do know that, given the relatively slow rate of biological evolution of species, the neural network of a human being will not undergo any fundamental reorganization in that time. Because society will consist of people very similar to us, we should be able, with the help of the general theory of sociology, if not to envision the *contents* of life in the future, then at least its possible *forms*, its possible structures, for the simple reason that it is the structure of human neural networks that affects social dynamics and determines the teleological, evaluative, symbolic, and self-preserving activities of the humans. The content of symbols, goals, and values of the future may differ as much from ours as ours differ from those of the Neanderthals, but the fundamental laws of the behavior of neural networks remain essentially the same. A social system whose members follow those laws—regardless of the motivational, economic, or metaeconomic specifics of their activity—must exhibit universal regularities that a theory can articulate.

Here is an example of how partial freedom from economic motivation may affect behavior. In societies with the highest levels of well-being, there appears the paradoxical phenomenon of rebellion "without a cause," in which groups of young people attempt to destroy the existing order and values purely for the sake of destruction. No other reason can be found. These youths are free of concern about their existence; they do not need to struggle for a career or livelihood. From the economic point of view, their behavior is irrational. But our theory predicts this, based on the general characteristics of human neural networks, as a manifestation of active teleological behavior and creative valuation when the economic motivation for action is weakened. In a society with a sufficient level of well-being and established general order, an individual who longs for a goal not preset by social conventions and seeks new values (most often it is young people who do this) must possess certain intellectual abilities. Without them the individual is unable to join the ranks of those who create new material and spiritual goods and can therefore externalize this desire only by opposing what exists. I am simplifying, of course, and a psychological analysis of these occurrences might explain more—but in our theory, which is by nature statistical, that explanation would be irrelevant. My point is that it will be possible to analyze such general regularities by formal, mathematical means—through studying the dynamic osculation, or the absence of it, between the individual processes in the network and the overall processes in the system.[5] Since the number of formal processes (or the number of possible variants in connectivity and excitatory/inhibitory influences) in a neural network is incomparably smaller than the number of the contents variants in those processes and since the number of possible, and stable, social structures is likewise smaller than the number of all processes that can possibly take place in them, solving this task should not be as difficult as one might expect based on traditional psychology or the traditional treatment of the relation between an individual and society. Now, perhaps, you will finally allow me to return to the topic, which is, frankly, more modest than these exalting but unavoidably too general divagations.

HYLAS All right, but only after I say what I have to say. I think I understand what you want to convey. The desires, hopes, drives, fears, and pleasures of an individual enter this world without clothing and without direction. It is only the existing reality that gives them shape, form, direction, and a plan for action; brings to life all the forces of the human soul, which do not know themselves; and gives them names and meanings by incorporating them in the order of social structure. I can imagine the language, taken from triumphant physics, of the future theory: the social continuum, probabilities, potentials of sociodynamic transformation, predictions based on statistics, and the deindividualized, mass nature of those transformations. Just as the physical state of a cloud of atoms with various velocities is determined by external conditions—the walls of its container and the pressure exerted by a piston—and just as the action of those atoms depends on the shape of the channel into which they are funneled—so that they move a turbine or expand while cooling or crystallize into ice under pressure—the social atoms subjected to various mental temperatures and pressures will do work that is determined by the dynamic structure of their ensemble, which is, in turn, predetermined by sociological engineers. All possible variants must be considered so that the parameters do not "misbehave," so that any possibility of initiating a chain reaction that could destroy the structure or form dangerous local condensations or dilutions is ruled out, as is the possibility that the "atoms" in the absence of a regulating channel use their natural energy for nonconstructive acts of destruction without any rational incentives or justifications. I understand the necessity of such foresight and action and applaud the resolve to avoid the use of force by replacing it with the very links that connect the self-organizing social atoms and whose objective emergence under any given conditions and parameter values can be predicted by a mathematical theory, in which energy that is exchanged and transported in collisions between gas molecules will be replaced in the social system with information charges that propagate from one individual to another.

I see the inevitability of this, but why, instead of celebrating this future triumph of human beings over themselves, do I feel anxiety? Excuse my weakness, Philonous, for solitary lyrical ponderings under the stars. I think of Dostoevsky and his words from *Notes from Underground*: "There will someday be discovered the laws of our so-called free will—so, joking aside, there may one day be something like a table constructed of them, so that we really shall choose in accordance with it. . . . You tell me again that an enlightened and developed man, such, in short, as the future man will be, cannot consciously desire anything disadvantageous to himself, that this can be proved mathematically. I thoroughly agree, it can—by mathematics. But I repeat for the hundredth time, there is one case, one only, when man may consciously, purposely, desire what is injurious to himself, what is stupid, very stupid—simply in order to *have the right* to desire for himself even what is very stupid and not to be bound by an obligation to desire only what is sensible. . . . It is just his fantastic dreams, his vulgar folly that he will desire to retain, simply in order to prove to himself—as though that were so necessary—that men still are men and not the keys of a piano, which the laws of nature threaten to control so completely that soon one will be able to desire nothing but by the calendar. . . . If you say that all this, too, can be calculated and tabulated—chaos and darkness and curses, so that the mere possibility of calculating it all beforehand would stop it all, and reason would reassert itself, then man would purposely go mad in order to be rid of reason and gain his point!"[6] If Dostoevsky is right, then what of your sociodynamic theory of the golden age, Philonous?

PHILONOUS Yes, he is right, and this truth can be—oh, paradox—expressed in mathematical terms, as a prediction of the constant appearance in a community of nonconformist individuals, both positive (creators of values) and negative (destroyers of values). But what exactly is Dostoevsky opposing so forcefully? The predictability of human actions, if I understand correctly. Yet a precise, deterministic predictability of human behavior is a fiction. We can calculate probabilities, not predict specific actions; if someone acted in certain way just to defy the "psychological forecast," he would only have my sympathy. Nevertheless, I suspect that your point is not this but the "crystal palace" of the future society, as Dostoevsky called it, or the "ideal system," as we did. As for life in this "palace," it does not mean uniformity or equalization—on the contrary! The dependence of one individual on others, which is inescapable in any social structure, can indeed lead to harmful stratification, to the exclusion of some groups and the elevation of others. But we can also set up such a distribution of fields of individual activity that, if we focus just on two people, person 1 will be subordinate to person 2 in domain A, while in domain B person 2 will be subordinate to person 1. This way, a relative equilibrium between dominance and submission would establish itself and be distributed evenly throughout the dynamic structure of interpersonal relations, which would prevent the emergence of any harmful automatic mechanisms of hierarchical stratification of society.

Such stratification arises not only from direct relations between two or more people, which we might call "short links," but also from the "long links," which represent institutional assemblies that disturb the homogeneity of the ensemble. The latter type of links set themselves apart by profession, for example. Economically determined relations, such as ownership versus lack of ownership of the means of production, also are "long links." To strike the right balance between the segregating and stratifying tendencies of the long vs. short links is the touchstone of any theory of sociological engineering and at the same time its most difficult challenge. If we do not wish (and we do not indeed!) to stabilize the ideal structure by force, which would lead to uniformity and a reduction in complexity, then we have no choice but to make the interpersonal relations and the whole structure more complex, of course, not just randomly but according to the theory's guidelines. In a structure that is dynamically stabilized by the overall formal complexity, no segregation (based on dominance-submission relations, on mental or physical traits, etc.) will emerge and a greater variety of options will open for an individual than in any structure known in the history. There is no reason here to worry about social determinism. But the issue of "limiting the moral costs" of social experimentation, without which any future progress will be impossible, is important. So all that remains of Dostoevsky's argument is a deep-rooted opposition to any kind of social modeling or engineering as a matter of principle. Which amounts to a surrender to the spontaneous and blind mechanism of the existing laws and a passive acceptance of all their harmful consequences—for they exist whether we like it or not. By analogy one might feel disgust for any sort of "artificial" treatment of a disease, advocating instead a reliance on the "natural powers" of the human body. The consequences would be most unfortunate. All the evils that humankind has ever suffered, and is still suffering today, as a consequence of our being an *animal sociale*, were caused by a lack of knowledge or by knowledge applied imperfectly or incorrectly, but never by too much knowledge. With this, I would like to conclude the argument about these *imponderabilia*, generalized to the highest possible degree.

We were talking about psychometrics. For the future sociologist-engineer it will be a helping tool only, never a recipe for an action, because which positions need to be filled, which professions and functions are socially in demand, what the organizational plan for production and distribution should look like—all this is determined not by psychometrics but by social structure (plus the level of civilization). Psychometrics, like material science in engineering, can tell us with what (or more properly, with whom) to fill the vacant spots in the construct, but the construct, the blueprint, must already exist.

Let us now return to our point of departure, that is, to the importance of intelligence and the role of intellectuals in a society, about which we can say the following: the future path of humanity's development does depend to a large extent on intellectuals, people with higher mental abilities than the rest, but the cohesive functioning of the social machine, without which a society here and now cannot exist and therefore evolve, is determined mainly by the noncreative—or, better, less creative—work of the majority of people who are average; this way mental skills and abilities of all members of society complement each other. Psychometrics, as a differentiating measuring tool, should only serve one purpose, to make sure that the right people are directed to the right positions in the system; under no circumstances should it turn into a tool for hierarchization, which introduces elitism, social inequality, and privilege.

Please note that the creation of thinking machines introduces a new kind of equality, because there is no longer any human activity that in principle cannot be performed by a machine. Not that long ago people thought that only "lower" activities, such as physical labor, were subject to mechanization, that mental work could never be. Now the line between mental and physical labor is disappearing, which means that the tendency of intellectuals to consider themselves the most valuable members of society must soon disappear as well.

HYLAS   It really wasn't my intention to pit intellectuals against physical laborers. I was only protesting against stupidity. We can safely assume that there are stupid shoemakers, but there are stupid intellectuals as well, who can actually cause more harm than the stupid shoemakers.

PHILONOUS   Even the stupid should not be judged hastily and denied their right to coexist with other people, for we all must live on the same planet. Any attempt to wipe out stupidity as a social phenomenon inevitably leads to societal stratification, until the day arrives when we acquire the means for the biological transformation of our species. In every society there are people on the low side of the normal Gaussian distribution curve of intelligence, and they are an integral part of society. To isolate them, whether intentionally or not, will only lead to additional stratification, which, just like the stratifications by race or class in the past, cannot bring humanity anything but suffering and war. Not even our ideal system can make incompetent people competent, and to expect that of any sociological construct would be a dangerous utopia out of touch with the real world. People who differ from one another in personality, mental speed, character, and intelligence—in short, all people—must coexist within the same social system. Because they differ, their suitability for various occupations will also differ. It will be the task of psychometrics to point them to paths that give them the most satisfaction from their activity and at the same time ensure the optimal fulfillment of society's needs; the selection should have nothing to do with an evaluative stratification of society.[7]

Apart from the stratification resulting from evaluative differentiation based on mental (intelligence, character) or physical (race) traits and apart from the traditional segregation based on birth, pedigree, and so on, there is the danger of stratification according to economic factors (the main subject of Marx's theory). In the highly developed societies of the future there is also the possibility of quasi-caste segregation due to professional and scientific overspecialization.

Predicting and combating these phenomena is the first duty of the engineer-sociologist. Because a society must be cohesive, we cannot be too careful in our projecting and constructing it. As we are well aware, those put in charge of the feedback between the government and the economy tend to elitistically monopolize their function, particularly if it is supported by a centralist feedback system. It is true that harmful tendencies toward stratification, division, and eventually antagonism between groups and their interests, which are sources of conflicts, appear in all kinds of systems. But it is also true that our ability to experiment, search, and choose in this area is limitless. Despite all the disappointments, defeats, and tragic mistakes of the past, people will build a better world. Should we act without this principle in mind, we would lose faith in humanity and its potential, and in that case, my friend, it would be better not to live.

Kraków, 1954/1955–December 1956

SUPPLEMENT 1:
THE *DIALOGUES* SIXTEEN YEARS LATER

LOST ILLUSIONS, OR FROM INTELLECTRONICS TO INFORMATICS

1

The disillusion in cybernetics in the two decades since its birth has been due to both practical and theoretical reasons. Since the theoretical reasons are more fundamental and also more difficult to articulate, I start with them. True, the "fathers" of cybernetics—Wiener, Shannon, and von Neumann—at the very beginning warned against the excessive optimism of treating cybernetics as a universal key to knowledge. But they themselves could not always avoid slipping into just such optimism.

The maximum epistemological program of cybernetics declared—nota bene more in popular accounts than in the scientific literature—that a new language, a new system of abstraction, and a new level of generalization had risen, allowing for the unification of the natural sciences and humanities, which until now had been separated by an impassable barrier (biology, geology, and physics on one side; anthropology, psychology, linguistics, sociology, and even literary studies on the other). Cybernetics could accomplish this because it had at its disposal models with such a high level of abstraction that they applied to a great variety of phenomena in the diverse sciences while they still preserved the identity of its central concepts—information, its sender, its receiver, and its transmission channel; a system equipped with "inputs" and "outputs"; negative and positive feedback; system trajectories determined by transformation matrices; and so on. These mathematically defined concepts were supposed to become the common denominator of all disciplines, and they would allow rigorous research in areas that had not been accessible to exact methods.

This promise was not fully realizable from the start, for two reasons. The first relates to the inadequacies of cybernetics itself—about which later. The second has to do with the difficulties that have plagued twentieth-century mathematics. Because historically mathematics privileged and adopted into its methodological arsenal disciplines that did not come from the mainstream of pure mathematics but developed somehow alongside it, such as the theory of probability, the basis of Shannon's information theory, which the pure mathematicians treated for a long time like a stepchild, and the theories of algorithms and of systems, which, not having achieved full formalism, operate with concepts that a mathematician regards with suspicion. Lacking the expert knowledge, we may resort to a metaphor: the mathematical foundation on which cybernetics was set was not completely solid. Probability theory and algorithm theory (in contrast with systems theory, so important for cybernetics, which is still more wishful thinking or a set of sketches and propositions than a true self-standing discipline),[1] both possess clearly defined centers and peripheries swarming with unresolved questions and doubts. Attempts to expand the centers have led to enormous difficulties that relate to the ambiguity of the two central terms, "probability" and "algorithm." In both these disciplines we can do either very little but with absolute certainty or a great deal but with little certainty. Yet using just the theories of probability and algorithms was insufficient for developing cybernetics, and it was expected that pure mathematics would provide a helping hand; among others, von Neumann expressed such a hope, seeing the inadequacy of the solid but not sufficiently powerful combinatorial procedures stemming from Boolean algebra, which had been resurrected and was suddenly fashionable.[2] Unfortunately, help did not come. Besides, mathematics had a hovering cloud of its own curses to contend with, related to infinity, because physics, especially classical physics, was based on infinitesimal calculus, but an infinitesimal apparatus embedded in set theory is of no use to cybernetics. This clash between the finite and the infinitesimal has led to fundamental questions about the nature of cybernetics. Is it a branch of pure mathematics, or is it a mathematically interpreted modeling arena of physics?

This question is not as speculatively barren as it may seem. As a part of physics, cybernetics would have an empirical origin and, constituting a theory or set of theories, it would necessarily have to be subject to experimental testing. As a part of mathematics, cybernetics would be a generator of models and structures that are true by definition provided that they were formed in a consistent manner and the question of the epistemological fruitfulness of its applications outside mathematics would be a separate issue, rather peripheral to a cyberneticist. We note that the cyberneticists themselves could not make up their minds regarding the status of their discipline. A Freudian historian of science might say that they were not willing to give up either the deductive truthfulness that is the hallmark of mathematics or the instrumental fruitfulness that characterizes natural sciences, and thus it was their "subconscious" that rendered the act of classification difficult for them. The "fathers" were all mathematicians by education, but only Wiener was close to pure mathematics; von Neumann was involved in a little of everything—quantum mechanics, the theories of automata and information, chemistry, biology, and even neurophysiology; while Shannon was a communication engineer.

The links were unclear, and still are, between cybernetics in the exact sense—which includes the theory of systems that possess input/output and feedback, along with all possible variations in the areas of homeostasis, self-organization, etc.—and the relatively independent fields, such as Shannon's information theory, and those with even looser ties to cybernetics, such as the theories of dynamic programming, decision-making, and organization. From the beginning, the main criticism of cybernetics was that it did not uncover anything new but only translated into its own language the systems and processes that were already well described in other languages; that cybernetics was therefore sterile. It is true that in many disciplines the conceptual apparatus of cybernetics proved unproductive. Cybernetics explains a lot—for example, in theoretical biology—but led to no substantial discoveries. Not that its applications were wrong; they were only sometimes premature and sometimes unimportant, when, for instance, the dearth of appropriate objective data in a given science made it impossible to fully flesh out a newly introduced cybernetic model.

It later turned out that information theory was inadequate to have the universal application that had been postulated with so much fanfare and that the notion of information, even the asemantic information of the Shannon type, was extremely difficult to use rigorously beyond human-made communication systems. I will devote a few words to this important issue. The developmental problems of cybernetics were not limited to information concepts, essential as they were, because it was hard to talk with rigor about regulation and communication in a system or a machine without having adequate tools at hand for reliable measurement of this communication and regulation along with the information involved.

The greatest joy, almost as if at the discovery of a modern philosopher's stone, was elicited by the equating of the communications information with thermodynamic entropy, which, as the "fathers" (e.g., von Neumann) claimed, built a bridge between logic and physics—for the first time in the history of knowledge. It was discovered that information is "negative entropy" or "negentropy," the reverse of entropy as a physical quantity that measures the "energy deterioration" or probabilistically understood "degree of disorder" in a system (because the highest order is always thermodynamically the least probable state of a physical system). But the joy quickly faded when it turned out that the optimism had been greatly exaggerated.

At first it was thought that the inadequacy of the physicalist concept of information was due to its asemanticity or the fact that Shannon's theory did not consider the content of information. There were attempts to construct a semantic information theory (e.g., Carnap and Bar Hillel).[3] But it quickly became clear that even the purely orthodox theory of information, the communications type derived from thermodynamics, suffered from inadequacies. This was discovered when researchers in different fields tried to estimate the information contained in a living organism, in an egg, a chromosome or gene set of a biocenotic population,[4] or just to determine which contains more information, a zygote or the organism that arises from it. One of the most brilliant linguists, logicians, and informaticians (all at the same time), Yehoshua Bar Hillel, eventually showed that questions in the form "What is x?" (i.e., "What is information?") are not permitted in science, that they have a metaphysical bent, as it were: they assume an "ultimate" answer, which would provide access to the "nature" of an "entity" like information or gravitation, to something permanent and unchangeable, something that science, constantly self-correcting and evolving, can never attain. His postulate has precedents. For example, there exist questions in quantum mechanics that are prima facie legitimate—that is, logically correct—but must not be asked nevertheless, as the answers to them would be self-contradictory and therefore worthless.

The radicalism of Bar Hillel's view unavoidably brings about crushing consequences: we cannot know what information really is, whether it belongs to the category of concepts like energy and mass or is a "special entity"; worse, we are losing the sense of the scope of its applicability, because if information, operationally undefined, is not a reasonable measure of anything in biology or psychology, then it cannot be used beyond the narrow domain of communications engineering technology, that is, transmitters, channels, and receivers, except in linguistics. Linguistics is saved for informatics because it can be treated purely combinatorially and probabilistically at the same time. Yet a full mathematization of linguistics is very difficult if not impossible, so we must satisfy ourselves with a heuristically legitimate approximation: for example, if we want to measure the frequency of certain letters in a natural language, we can obtain an answer with an arbitrary degree of precision but without a closed theoretical formula, because the formula depends on observing the speakers of the language, and we are back at the good old dichotomy between empirical probability and the mathematical expectation theory.

But limiting the informaticists, if they are linguists, to the study of natural languages would be damaging, as it would cut off any chance of a parallel investigation of that "other" language, the genetic code in biology. The empirically evident similarity between genetic messages and language utterances raised great epistemological hopes. It seemed possible to achieve a new type of generalization, in a synthesis of which the concept of information would play the leading role. It seemed that information, originating simultaneously from the fields of logic and thermodynamics, would enable a unified approach to both natural languages, which serve interpersonal communication, and effectory languages, i.e., self-realizing prognoses that are the chromosomal articulations of living organisms. To the empirical natural scientists, everything pointed to the common roots of these two classes of language, but the theories that were supposed to underpin this happy synthesis and open the door to future research proved to be too weak.

The time bomb thrown into this field of inquiry was the concept of complexity. The whole theory of systems is being built just to clarify this foggy notion which refused to submit to any precisification, especially of the mathematical kind. There were attempts to bridge the concepts of physical information and complexity so that a measure of information would automatically give the degree of complexity of the object being studied. Relics of those attempts can be found in terms like structural, metric, topological, and algorithmic information, which go beyond probability and combinatorics. Unfortunately, they all brought more disappointment than success. Let me explain in a brief, accessible, and therefore inevitably sketchy manner why the notions of information and complexity were incompatible.

To many, information suggests a subjectivity that is foreign to physics. The measure of information is always contingent on a repertoire of states: receiving information is equivalent to the act of choosing one state out of all possible states, which is why it is measured as a before-and-after difference in probabilities when a signal is received. The signal's arrival is equivalent to a measurable decrease in the uncertainty of the state, which the observed system actually exhibits. The system is therefore treated as an element of a finite and countable set of all possible states. For the linguist, there is nothing simpler than determining this set's boundaries, because its elements are language utterances, which are always finite, and the amount of information can be determined exactly using combinatorics based on the number of signs (letters of the alphabet) that appear in the communication. Simple combinatorial operations allow us to perform the measurement. Whether the words are written horizontally or vertically, on a plane or the surface of a sphere, or whether they are encoded into a material substrate such as paper or transmitted as phonemes through oscillations in the atmosphere, makes no difference. The amount of information in a printed sentence does not change when it is written with a stick in the sand or when the letters are shaped from clay, cast in metal, or carved into a stone. But in the case of the genetic code, we have no idea how to separate the information from its substrate. A set of sentences in English and a set of "chromosomal sentences" are not amenable to the same accounting with regard to information content, because the relation between the emitter (the information source) and the receptor (receiver) is drastically different in the two. We know what the set of English sentences is, but, although we can assign the specific units of the genetic code—DNA—to specific letters of the Latin alphabet, we don't know what exactly constitutes "a set of chromosomal sentences." We can distinguish between a bunch of letters and an English sentence but not between a random chain of "gene letters" and a "sentence" in the language of heredity. Yet our current inability to do so is not the only serious obstacle on the path to measuring information, for the reason that sentences in a natural language are never true systems in the physical sense. The systemic nature of a language sentence is not its physicality but its congruence with the rules of syntax, lexicography, and grammar. If we were to treat a printed sentence as a physical system and wanted to measure its thermodynamically understood information, we would find that that the amount of order that the print contributes to the total entropic balance of the page is so minuscule as to be practically zero. From the thermodynamic point of view, the page sprinkled with printer's ink and the page with a meaningful text printed on it are informationally, that is, entropically, almost identical. The reason is that one bit of information corresponds to 10^{-16} thermodynamic units of entropy.[5] The amount of order in a physical system such as a sheet of paper is practically unchanged when we print letters on it because the text may contain at most a few hundred bits of information, whereas the entropy of the sheet transformed into bits is astronomically large (hexillions or quintillions of bits). The thermodynamic-informational balance in any human-made object, for example, a digital machine with 100 million elements, which can contain at most 4×10^9 bits corresponding to 4×10^{-7} thermodynamic units, is again insignificant. The situation changes when we consider a living system, whether it is a mature organism or just a germ cell, because the number of elements constituting such a system is on the order of 10^{12}; therefore, the *physical entropy* of a cell already depends on the *amount of information* that it contains and that governs its behavior (e.g., the process of embryogenesis).

But insurmountable problems arise the moment we attempt to determine the amount of information in the chromosomal fiber because the simple statistical procedure that we employed on the sentence composed of letters cannot be applied to the "sentence" made of the units of DNA (deoxyribonucleic acid). The topological properties of the printed sentence do not affect its information content, but the topology of the chromosomal fiber does. This is why we cannot equate the DNA units, that is, the genetic alphabet, with a linguistic alphabet. We should realize that the ease of the procedure inherent to the informatic treatment of language texts is simply the consequence of omitting all the connections that the perceived sentence makes with the perceiver's brain. The sentence turns into a system of commands (directives) and, at the same time, into a substrate of a complex decision-making procedure in the brain, just as the chromosomal fiber is simultaneously a system of commands addressed to the cytoplasm of the zygote and, along with the cytoplasm, a substrate of the complex decision-making procedure that is embryonic development. But sentences of a natural language are always considered completely separate from all the operations of their interpretative reception, while sentences in the language of genetics do not exhibit such autonomy. Although we believe that this difference is not fundamental but due to technical aspects of the communication, it is precisely these technical aspects that prevent us from using here the simple methods of statistics and probability theory, which suffice for information accounting in linguistics. Biologists who repeatedly attempted to determine the information balance in cells and organisms ended up with wildly scattered results, and, worse, with fundamental errors and misunderstandings in the concept of information, which often lost any physical, operationally justifiable meaning in the process.

For example, it appears that the amount of information in a living organism is the same or almost the same as when it is dead and that the amount of information in a zygote is smaller than in the mature organism that develops from it—which would justify the assertion that life goes "against" the entropic gradient and is therefore not subject to the laws of thermodynamics. These undoubtedly false claims are based on faulty reasoning that is in contradiction with physics. We simply cannot investigate the processes of life with the same strong limitations that are acceptable in the study of language sentences, which are isolates. What can be omitted in linguistics cannot be omitted in biology; doing so would lead to nonsense. When Shannon asked what he should call the fundamental quantity in his theory, von Neumann suggested entropy, not only because of the mathematically identical formulas but also because, as he mischievously quipped, nobody really knew what entropy was. When several years later Leon Brillouin was writing his book about information in science,[6] he called entropy the measure of our knowledge about a physical system—not simply the measure of "disorder" in the system. This elicited protests and misunderstandings; many thought that Brillouin considered entropy, and therefore also information, a purely subjective measure, indicating what we know about the system rather than marking an objective feature of its state. So von Neumann was right.

The notion of entropy is derived from investigations of various states of a gas, particularly the perfect gas, and statistical mechanics; its magnitude expresses our knowledge about a system in the sense that the system, e.g., the gas in a container, can exist in one of the innumerably many states that we are unable to distinguish because we cannot determine the positions of all the gas molecules at the same time. Therefore, entropy relates to all these indistinguishable states *taken together* at a given moment; because they are all equally probable, from the "entropic point of view" they are equivalent to a single macroscopic state. Hence, entropy indeed relates to subjective knowledge but coupled with a rigorously objective state of the system; because when the system transitions into states that are progressively less probable and more ordered, its entropy decreases, *and thus the number of different but equivalent configurations of gas molecules must decrease*, even though we cannot measure it. Zero entropy, unattainable in principle (since uncertainty is unavoidable at the quantum level), would represent a state in which we know everything about the system (in a physical, i.e., spatial, and not metaphysical sense) because it has reached a completely determined, and what's more, *the only possible*, configuration of molecules.

Chromosomes are highly ordered systems, since every particle in them must occupy a given position; their entropy is very small and their information content enormous, possibly approaching the maximum for a polymeric macromolecule. As we said, the order of the letters in a printed sentence in a natural language is thermodynamically almost zero compared with that of a "genetic sentence." The communication efficiency of a natural language is a consequence of its "triggering" character because in the case of language articulations, the brain acts as an enormously powerful amplifier on the physical level: a sentence that has been uttered and heard and whose entropic content is thermodynamically extremely small sets off in the brain an avalanche of coordinated processes that enable the "understanding" of the minuscule portion of entropy that was transmitted. Thus the sentence acts as a cock or trigger that initiates this multicascade amplification also in terms of energy: the "understanding" of the sentence by the brain requires an amount of energy that is gigantic compared with the thermodynamic balance of the sentence itself, although still small in the absolute terms—the total power of the brain is ten to twenty watts. A "sentence" made of genes, however, is not merely a triggering device but an autoliberator or self-effector that initiates, organizes, and regulates the whole process of embryonic development, which would not be possible if the chromosomal fiber did not possess an extremely high degree of order at the start. Linguistic entropy is not thermodynamic entropy, *because the systems studied by linguists are not physical*: the embodiment of an utterance plays no part in the processes of language communication understood as information transmission. Sentences of a language are primers that trigger a highly ordered avalanche of brain processes, and the process of reception gradually acquires physical character as the physical aspects of the brain functioning cannot be ignored, in contrast with the physical aspects of the language articulation itself (i.e., whether it is printed, sculpted, engraved, etc.). This allows the use of simple, almost primitive measuring methods in linguistics, whereas similar measurements in genetics are enormously difficult. The reason is that the *physical* aspect of a genetic "sentence" can *never* be omitted in the determination of its informational content. That is also why the pure combinatorics as a part of logical analysis suffices in linguistics but is of no use in informationally understood genetics: the chemical, molecular, and quantum-topological aspects of a written sentence are irrelevant, but the chemical and quantum-topological aspects of a chromosome are the essential determinants of its order.

A sentence in a natural language is true if it is constructed according to the rules of lexicography, grammar, and syntax—though not necessarily of semantics, because the sentence "The safety pins spend the night unusually in the crater corkscrews" is linguistically correct although its meaning is questionable. A genetic "sentence" is "true" if it represents a system of directives aimed at achieving a certain final state of the system, which is a mature organism. This "sentence" cannot be syntactically correct but prognostically (or effectorily, that is, teleologically) incorrect since the syntax of the genetic code is embryogenesis. If a set of genes (DNA units) does not trigger embryogenesis, it is not considered a genome but merely a chain of DNA elements that is chemically possible but causatively, embryogenetically, barren. One could theoretically create all the combinatorially (and chemically) possible systems of DNA as a set of macromolecules with the size (length) of the real genomes found in nature. Their number would be on the order of 10^{3000}, meaning that there are not enough electrons in the universe to embody this set. But its technical impossibility aside, such endeavor would be meaningless; it is just to illustrate that the "genetic building blocks" can be combined even when they do not function as "genetic building blocks" at all. The point is that a biologist-geneticist does not want to measure "all the information" contained in genes (in molecules, atomic ensembles, electron clouds, quantum-mechanical systems, etc.) but only the part that actualizes embryogenesis. He is therefore interested not in "all bits" but only in the "biobits," that is, the regulatory quanta of embryogenesis. But these "biobits" have little to do with the entropy of the linguists or Shannon. This is the obstacle in the creation of a "generalized linguistics" whose special cases would be, on the one hand, all natural languages and, on the other, all genetic codes.

There were attempts to save the situation by altering the basic concepts of information theory, but all nonprobabilistic information theories share the disagreeable property of lacking an elegant, natural transition from the notion of information to the thermodynamic, physical notions of Shannon. At the same time, considering information to be an entity that is subjective in nature, an idea that some linguist-informaticists are favoring, is baseless: zygotes, embryogeneses, and transformations of genomes into organisms existed billions of years before the rise of human beings and their natural languages. It appears that the mathematical tools used to avert the crises are too simple (and theoretical biology cannot gain much from graph theory). I do not believe that pure mathematics is of any help, because it can never lead to the concept of a threshold of the minimum complexity of a system, which is essential to an understanding of the phenomena of life. So far our realization that the amount of information in an object is not a clearly defined function of its complexity is only intuitive, and attempts, such as those of Brillouin, to divide information into "free," that is, lacking physical interpretation, and "bound," that is, correctly reflecting the information content of an object in the physical sense, skirt the problem instead of solving it. Thus we only intuitively understand that a system capable of self-reproduction must exhibit a certain minimum complexity, below which it cannot function, regardless of its structure. (I touched on this matter, in a slightly different context, in an essay on value in biology, published in *Studia filozoficzne*. It is included in this book.)

Apart from the failures in the field of the theory of knowledge—with its program to unify the various natural sciences elevated to a cybernetic "metalevel"—cybernetics suffered other fiascos. Many people believe that the notion of Shannonian information has been left "unfinished," hanging in the air, as it were, while we must strive for a synthesis unifying it with other physical concepts, much as, *mutatis mutandis*, the theory of relativity unified time and space into a four-dimensional continuum. I fear such a hope is fundamentally wrongheaded, because nothing can be achieved on this terrain in a simple, clear, and at the same time precise (i.e., quantifiable) manner. This does not mean that I am pessimistic about the future of cybernetics;

I just do not think that any single terminological or conceptual invention or revolution will lead to a breakthrough that makes cybernetics an epistemological cornucopia and pays back with interest the debts that it incurred with its initial bold promises.

Yet other failures, more technological than theoretical, were the dashing of hopes for the construction of "an intelligence amplifier,"[7] a translating machine, and a machine that would, finally, imitate (even if only on the level of language) a human being (Turing's idea). I think we can attribute these disappointments to the difficulties, hidden at first, that the theory of automata or indeed all of computer technology had to face. Computer programmers encountered unexpected problems as the programs became more and more complicated. Neither the inadequacy of machine memory nor the uncertainty of general operational strategy (should computation be "parallel" or "serial"?) limited this field's achievements as drastically as the issue of program construction, which *must* become a part of a physically interpreted theory of algorithms. But this goal is too far, and enormous difficulties tower on the path to it. To put it in succinct, and therefore apodictic and simplifying, terms, the early optimism of the cyberneticists was based on the view, usually not expressed explicitly, that intelligence could be automatized by replacing mental processes, such as the ability to search, with *mindless* procedures by inserting the appropriate algorithms into a program that performed a task. A search for the best chess-playing program is nothing but an attempt to construct—with the method of successive approximations, trials, and errors—a fully functional approximation of the chess algorithm, which has so far not been fully achieved by means of pure mathematics. Also programs capable of learning were supposed to appear, except that their education would be implemented by another algorithm. And what was the solution finder in Ashby's intelligence amplifier if not an algorithm for filtering? The strategy has always been to compress the mechanism of any kind of reasoning into its minimum form: a recipe, as universal as possible, embedded into a structure of network connections. The underlying assumption was that much of the structure of the brain was redundant and could be omitted. But is not the reason for toning down our optimism staring us in the face? Because if any form of survival tactics in terrestrial environments were algorithmizable, evolution would have unfolded differently than it actually did. Bear in mind that evolution is quite an effective constructor, and the homeostatic products with which it populates terrestrial environments are very well adapted to their surroundings—within a given homeostatic plan. If any form of adaptive strategy algorithmization were possible, the set of those survival algorithms would form an attractor or sink in the phase space of evolution's speciation: because wherever an algorithm is in place and functions optimally to solve a particular problem, nothing that would work "even better" can arise. The algorithm for adaptation would be embedded in the organism's nervous system, and that would be the end of progress in the entire domain of life. Let us note that no species is "final" but each is a link in the creative chain of organizational solutions on the neural level that passes what is "better" homeostatically to the next link and also that a human being has such a large brain. These two facts support the thesis that it is impossible to reduce the survival heuristic to an algorithm, either as a mathematical formula or a trial-and-error method of successive approximations.[8]

The number of theoretical schemes proposed to represent the structure of the brain was unusual in the 1950s, but nobody tried to construct a minimum functional model based on them. How strange that this was the dream of the cyberneticists! Infatuated with the first step they had made on the path toward the imitation of thinking, they believed that thoughtless repeatable procedures could replace thought. They did not express it explicitly, but that was the goal they tried to reach—in vain. Were they wrong? Yes and no. Von Neumann, in his comparison of a brain and a digital machine, focused on differences in size and efficiency of the building blocks, because in the era of cathode tubes those differences were huge. Today, when we have monomolecular memory, transistors, neuristors, microminiaturized systems, and integrated circuits, the differences have disappeared—and yet we have come no closer to constructing "a brain substitute." The differences in size and information-transmission efficiency of the building blocks turned out to be unimportant!

Evolution seems to be a lazy constructor. It squeezes out of its products all it can, stubbornly sticking to a model already developed and repeating it in every possible way. It allows a radical change only once in a hundred million years or less. My point: if a system considerably simpler than the human brain could handle the problems that the human being encounters in his ecological niche, then his neural network would be *precisely* that simpler system. The celebrated, promoted, and condemned "redundancy" of the brain is fiction. The brain is redundant only in that despite the irreplaceable loss of about a hundred thousand neurons every day, it works fine in old age, when people have only 60 to 70 percent of their initial neural power. It is redundant only in the sense that it can cope with the loss of its mass, not in the sense that it contains reserves that are never active but whose activation would significantly and immediately increase a person's intelligence. The brain is constructed according to the same praxeological rule that we see at work throughout bioevolution. Its redundancy is an illusion that stems from our inability to comprehend how an organ that arose by natural selection to function in the time of primary anthropogenesis, in the Eolithic, and was adequate for the "cave dweller" level of problem-solving could be adequate also for the tasks of subsequent human history, from the construction of the pyramids and epicycles to the creation of the theory of relativity and computers—without a latent redundancy in the "cave" era. We are equally unable to comprehend how the mechanism that regulated the reproduction of microbes, amoebas, and trilobites could, basically in unchanged form and a billion years later, create dinosaurs, whales, Pliopithecus, and eventually human beings, since the genetic code arose just once, and its lexicography, syntax, and grammar are shared by everything that has ever lived on Earth. Thus the brain that was selected for "cave-type" tasks proved able to tackle tensor algebra and group theory. It has not changed at all; what has changed—culturally, not genetically—are the canons of its specific *programming*.

If the algorithmization of the epistemic heuristic were possible and intelligence could be automated, systems whose complexity is that of the human brain have not found the way. Such algorithms, if they existed, could be actualized only by systems more complex than the human brain, and this is precisely why evolution has not realized them. In its progress, evolution always chooses and solves the problems that are *easier*. So it seems that either such systems cannot be constructed at all or they can be, but the cost of their construction, in terms of time and the amount of creative combinatorics, is higher than the investment that the evolutionary process spent on us. There exists a third possibility, that the structural-functional system that our brain inherited from the hominid species, is grossly incompatible with the blueprint for an automaton of algorithmic gnosis; in that case evolution would again be unable to construct such automaton since it realizes changes not in big jumps but through the slow accumulation of small alterations. Note that the last possibility is not very probable because in the terrestrial environment intelligence is a value that almost universally facilitates survival and therefore every species having it would be selectively privileged—yet only the hominid group formed intelligence, which means that this terrain cannot be reached by taking a shortcut or breaking through a barrier. Considering the number of all animal species equipped with a nervous system that ever lived on this planet in the last few hundred million years, it sure looks like a shortcut to intelligence, understood here as an automatizing simplification of its acquisition through repeatable algorithmic procedures, simply does not exist. Some engineers who constructed otherwise original and logically valuable models of logical networks believe that such a path "must" exist, but the failure of their search is an indication that it does not.

Obviously we are wise in hindsight; sixteen years ago, the matter was less clear. In particular it was not understood why the human brain at birth had so little in the way of preprogramming, why a person had to learn from scratch practically "everything," even sensorimotor coordination. More hardwiring would have made adaptation to the world considerably more economical. At present we assume that the brain is negligibly preprogrammed precisely because too much genetic hardwiring would greatly diminish our chances to adapt and therefore survive. The reason is that making the brain is just one facet of the problem called the "acquisition of intelligence"—the other, separate and huge, is the appropriate programming of it. It is thus for good and strategic reasons that our brains are initially "underprogrammed," and their enormous plasticity and potential, in all known cultures in history, gave way to specific actual realizations, reflecting the universal fact that a hasty automatization of epistemic procedures always does more harm than good. Our world appears to be a place where an acquisition of a closed set of directives for universal epistemic effectiveness is either completely impossible or is possible only after scaling a barrier that is higher than the one on the path leading to the brain of *Homo sapiens*. In this light it is easy to understand the failure of cybernetics to show that what evolution has accomplished in a complicated and arduous manner can be achieved in a relatively easy and straightforward way. Clearly, if a universal algorithm for constructing programs of gnosis, reflection, and heuristics—understood as games played with Nature—could be created at little cost, it would prove the nonadaptive redundancy of the human brain. But the brain turned out to be not only a device much more complex than the experts imagined but also, and more importantly, a device whose "redundancy" with respect to its functions is very little or perhaps nonexistent.

This conclusion, along with the observation about the almost zero preprogramming of the human brain, makes me pessimistic about the possibility of our discovering simple and robust procedures that can be copied and multiplied, perhaps derived from algorithmic set theory, for the self-programming of systems of the digital machine type (i.e., the technological realization of Turing's universal automaton).[9] The situation reminds me of Einstein's comment that "raffiniert ist der Herrgott, aber boshaft ist Er nicht."[10] On the one hand, there is no doubt now that a device, so unexpectedly elementary as Turing's machine, can execute any operation that an arbitrarily complex structure such as a brain or even superbrain can do, which suggests that "der Herrgot is nicht boshaft." On the other hand, to do "everything" this simple machine must have programs that *cannot* be reduced to a single common denominator—an algorithm, which shows the "subtleness of the Lord," who giveth with one hand and taketh away with the other: the effectory apparatus and the substrate for thinking is simple, but programming it is neither simple nor universal . . . Thus the failures of cybernetics, which promised to construct an intelligence amplifier, an imitation of a human being, or a translation automaton, truthfully reflect the situation, in that these failures are just its technological and engineering consequences.

It will not be out of place here to turn to the inventions that natural evolution can boast in contrast with our many spurious inventions achieved *viribus unitis*.[11] Ashby's information generator, Chomsky's generative grammars, and my idea of "information breeding" all share a common feature: they generate diversity by broadly sketching theoretical and technological aspects of the respective creative action, but at the same time they omit or mischaracterize, in just a few general words and optimistic allusions, the related problem of the *selector* of this diversity. For what is the use of creating an abundance of articulations, concepts, theories, and structures when we do not know what takes the place of that part of the mind whose function is to *sift through* the alternative possibilities? What is the use of creating abundance when we have no idea how to find in it the tiny, microscopic fraction of structures that have value—meaningful sentences in the case of linguistics, rational thoughts in the case of the "intelligence amplifier," or sensible theories in the case of my "information breeding"? In each case the easy part of the task is solved and the difficult part is flippantly tossed to others to deal with. Whereas natural evolution created not only a diversity generator, which is the "articulation field of genomes," that is, the set of all genetic codes at the disposal of a population of all living individuals, but also a selector that preserves only what proved useful—the process of natural selection with a Markovian character.[12] This amazingly efficient two-part mechanism is a nightmare for human constructors, because the part that decides what has succeeded and what has failed—the selection filter—requires, in its evolutionary issue, *millions of years* to fully manifest its creative potential. Unfortunately, this is a parameter that we can never adopt from the evolutionary version along with all the riches of its inventory. We might accelerate the selection process a millionfold by outsourcing the job to "luminal" digital machines that work at the speed of light, but no matter how promising this prospect may seem, we do not know if modeling evolution with the necessary degree of complexity can be realized. The path to the goal might lead through building a kind of "evolutionary ladder"—a hierarchy of automata and procedures—such that simpler programs would aid in assembling more complex ones until after many stages systems appear that can outperform bioevolution, and not only in speed. But this path is just a hypothesis that refers to a very distant future, from which we are separated by a space of many unknown discoveries and revelations, which will bring successes, yes, but also many disappointments. And should the ultimate success in this evolutionary competition with Nature turn out to be impossible, it will mean that Einstein erred, and "der Herrgott" is not only "raffiniert" but mighty malicious too.

2

The divergence between the expectations and accomplishments of cybernetics raises the following question: If we are building computers but cannot create functional simulators of the brain because computers are much easier to make than brains, then why did evolution, which always takes the easier path, choose the more difficult one? The answer is that we are building universal digital machines but not equally universal programs and we are fine with that, because we use computers to solve tasks that do not require their full operational autonomy. Evolution never faced this choice: its products, living systems, never surrendered their full operational autarky[13] in exchange for narrow specialization—with three exceptions: when cooperation as a form of specialization led to the emergence of a loose aggregate of homeostatic units, that is, a colony of living organisms (corals, anthills); when cells lost their universality in the formation of multicellular organisms; and when evolution developed parasitism and symbiosis. Aside from these three cases, characterized by *group* survival strategies, organisms had to tackle the problem of constructing a central nervous system, which is an equivalent of a universal informatic machine, and, *at the same time*, the problem of creating programs for it, which is why the two problems never arose separately. For this reason, evolution developed mixed tactics and strategies in its products; every organism, morphologically and functionally a sovereign unit as a player engaged in the game of survival against Nature, must have a full autarky, since naturally it could count on no help from outside, especially in the area of information. In contrast, computers without people are helpless. So evolution's task was from the start qualitatively different from the group of technologically conditioned tasks that have enabled us to build digital machines.

If a digital machine—or its biological and at the same time isomorphic equivalent—could have arisen in evolution and successfully coped with typically homeostatic tasks, there is no doubt that this would have happened during the billions of years of development in the biosphere. Evolution's stepwise character stems from the fact that the advantages and disadvantages of changes that increase an organism's complexity never balance exactly, that a system that grows and complicates its *soma*[14] and brain acquires new powers but also new weaknesses—at the same time. Statistically, the advantages slightly exceed the disadvantages—otherwise the transition from simplicity to complexity would quickly come to a halt. A bacterium is not more complex than a modern universal digital machine, but it is the bacterium, not the machine, that will survive when placed in an arbitrary environment because the machine is not a sovereign homeostat. It is in this respect that the nature of the tasks in the evolutionary flow determined the direction of evolution's construction efforts, which differ radically from those of ours in informatic technology. Computers' lack of autonomy, their dependence on people, refutes the idea that they may take over in some kind of "revolution" and rule humanity in a kind of "computerocracy." This idea is based on false historical analogies with the struggle for power, supremacy, or the circulation of elites, which do not apply to intellectronics.

Computers will not dominate us unless we allow them to do so. And there are two ways how it could happen: either intentionally, that is, through the construction of governing machines—but then the issue is sociologically and ethically trivial because it was the humans who decided to enthrone computers on the pinnacle of government—or unintentionally, when the system "people + computers" gradually acquires an unexpected and undesired dynamic characteristic. When we discuss the various possible futures, our imagination, conditioned by our history, has been limited to visions of the "intelligence amplifier," the "homunculus," the "electronic sage" or "electronic demon," the latter considered especially suitable for the role of a tyrannical ruler. Such models are naive and unrealistic. This does not mean that the cooperation of people and informatic machines is without danger. But the danger is completely devoid of the "personality" element: if an intellectronic system rules over us, it will not rule as a kind of simulated person with specific character traits. However, this only makes the danger far greater than if computers acquired "personality." Because if they did, at least one of the two sides—computers—would be acting with full awareness of what's happening. When someone, as a domineering personality, fights for power, he surely knows what he is doing; he is acting with intention and according to a plan, ethical or not, that makes sense and has a purpose, which we might even be able to discern. But if a gradual accretion of informatic machines and memory banks creates a governmental, continental, and eventually planetary computer network, which is the direction of the current development, the system composed of people and the said network may acquire a dynamic trajectory that does not coincide with the interest of our civilization. More specifically, the system might begin to drift. A large and highly complex system has an innumerable number of rules; when we are creating it, we see the benefits but not the unintended consequences.

The present problem of technology, historically the first, is the split of the global instrumental potential into two. From the beginning of civilization until today, we had only one type of technology: directed at the production of energy or things or the transportation of goods and people. Such technologies served us directly. Their growth has led to the loss of the biosphere's self-regulating equilibrium, owing to which the next wave of technologies will have the new, exclusive task of reestablishing and maintaining this fragile equilibrium. If the first-wave technologies served us directly, the second-wave will serve us only indirectly—working not to help us but to save the terrestrial environment as a whole.

A counterpart of the physical labor that technology performs—such as mining raw materials and then transporting and transforming them, shooting cosmic projectiles out of the well of gravity, heating the arctic and cooling the equatorial dwellings of human beings, and so on—is the informational labor, yet both are similar in that they lower the entropy in one place at the cost of increasing it in another.

This type of transaction is imposed upon us by the nature of the world in which we live. We stand at the threshold of informatic technology, which besides gifts that are beneficial can give us also gifts that are Danaian.[15] The socioeconomic symbiosis of people with machines may dissolve the boundary between who is the mover and who is the moved, between who rules and who is ruled. The very large and complex global network system that will arise must have a very complicated structure of its rules. We will build it gradually, with an eye to specific benefits that in practice can be discerned clearly and early on. But the system may have dynamic features that are hidden from us—because of their innate inaccessibility, not because of anyone's perfidy—and they may imperceptibly push it into a civilizational drift. I repeat, this would not be due to the "cryptocratic" behavior of any computer. I also ignore here the danger, so often discussed, of the loss of privacy when all, even the most intimate data about each person are stored in a machine memory: if this danger becomes real, ways can be found to remedy that. The machine-human symbiosis will be marked by contributions from each "side"—shared participation in decision-making, governing, and controlling. Yet the system as a whole may acquire a dynamic that is *not entirely* accessible to *either of the two sides*, because no system can describe or control itself completely: this principle is inviolable. A system, through observation and generalization, can uncover particular laws of its own operation but never *all* of them. That can be accomplished only by a higher-level system that has control over the first system. But initiating such control would be, in informatics, equivalent to the division now occurring in the technologies of labor, that is, the creation of informatics "of the second order" that would not serve us directly but supervise the symbiosis of people and machines so that its *equilibrium* would not enter any undesired *drift*. This act, however, leads to a fatal *regressus ad infinitum*: it would require in turn a supervisor of a next order that will control the controller in answer to the inevitable question, *Quis custodiet ipsos custodes?*[16]

Solving this problem obviously belongs to the far future, yet it is worth mentioning, because it indicates that the divergence between initial human expectation and final realization is indeed a constant in our history. The image of an infinite series of "informatic mirrors" as a control pyramid suspended above the civilization of the future is surely strange, as is the process of "reflecting" or representing the entire global activity for the purpose of achieving the optimum control, but it is also an ironic evocation of our ancient beliefs, such as the myths of higher powers who know everything about every human's life and to whom everything must bend a knee. If the infiniteness of the supervisory pyramid caricatures the role of God, the archangels, thrones, and the whole hierarchical rest of the celestial informatics, it does so unintentionally: such an infinite regress is impossible to realize, and therefore an informatic machine that would simulate God's controlling omnipresence will never come to pass.

The above prompts the following reflections:

(1) Research communities that face significant innovation tend to divide into two opposite camps, both armed with the slogan "Everything or Nothing." The adherents of cybernetics expected everything from it; the critics of cybernetics considered it epistemologically almost useless. The middle ground was much less common; I argued in its favor in my *Summa Technologiae* (1964), in the section "Doubts and Antinomies" of the chapter "Intellectronics." Attitudes of critics like M. Taube were outright liquidatory.[17] On the other hand, incurable optimists like to extend the deadlines by which "everything" will be accomplished—which is equally damaging for cybernetics—and some even falsify data. A book published in 1968 by Gallimard in the series "Idées," *Les ordinateurs—mythes et réalités* by J. M. Font and J. C. Quiniou, completely ignores the difficulties that various sections of the cybernetic program encountered, and even claims that in the USSR a novel by Dickens had been translated from English to Russian by a machine so well that the translation was equal in quality to a literary translation done by a human being (which is simply not true—Soviet sources are silent about it). Some professional logicians expressed faith in the unlimited power of cybernetics after Wang programmed a digital machine to prove most of the theorems in the fundamental work of Russell and Whitehead, *Principia Mathematica*, in 8.5 minutes,[18] a task that took human experts many *years*. Yet we still do not have a program that would enable an informal, inconsequential, friendly conversation

with a machine. This incongruity will make sense when we realize that although a human hand cannot lift a weight that for a crane is insignificant, it can perform thousands of operations beyond the ability of any crane. Our brain is ill-suited for narrowly deductive procedures, since it was constructed differently—to be versatile and universal. Proficiency in a language is determined by many areas, which simultaneously supply selectors that create an articulation and whose number, in rare cases such as a "difficult" literary text, can be innumerable. In principle, a machine is able to translate a terse scientific article with full proficiency even today if the author knows how a computer translation program works and is willing to write the article with that in mind. But no one would expect a scientist to go to such additional trouble—it might be easier for him to learn a foreign language to a necessary level than to write in conformity with the machine's translatory skills.

(2) Those opposed to cybernetics claim that a machine could equal a human being only if it were a living organism made in imitation, in other words a human being "created in a test tube" (e.g., the Dreyfus brothers).[19] Those defending cybernetics argue that the construction of an intelligent machine has failed only because of reasons that are beyond cybernetics: the enormous cost, the lack of market demand, technological difficulties, and so on. Those better informed know that these arguments do not conform with reality; the difficulties are principial and follow from theory. As for market demand, there obviously exist powerful groups interested in equipping the military with "intelligence amplifiers"—so the barrier to cybernetics is not economic. The persistence of enthusiasts has resulted in chess programs that can defeat any human player, but this success is a consequence merely of improvements in technical parameters of information processing, not of breaking through the barrier of heuristics or jumping to a higher level of computers' intellectual proficiency.

Both the apologists for and antagonists of cybernetics distort the reality. The optimists leaped at the possibility of bypassing all the steps that natural evolution took to construct us. They relied on the idea that the evolutionary process is equifinal with an algorithmic or heuristic procedure amenable to quick mechanization. But as the book

Artificial Intelligence through Simulated Evolution (1966), written by L. J. Fogel, A. J. Owens, and M. J. Walsh, showed, in an amazing confluence of arguments with my *Summa Technologiae* (1964, the first edition), many experts understand the necessity of going back and assuming a different, broader goal, that of modeling bioevolution as the "preceptor" of causative action also in the domain of *intelligence*. According to those authors, the modeling of evolutionary processes is indeed the first prerequisite for the automation of intellect.

(3) Finally, let me say it clearly: in the entire area of our discussion, the dichotomy is false between believing that a machine can be equal to a human being and believing the opposite, that humankind has eternal supremacy over all its creations in the intellectual domain. The optimization of machine parameters due to technological progress will not automatically take computer engineers across the threshold beyond which intelligent machines can be built. Nor is it true that only a synthetically replicated human being can equal a natural member of *Homo sapiens*. The task has turned out to be *several orders of magnitude* more difficult than it seemed twenty years ago—but no one has proved that it is unsolvable. Machines today easily cope with tasks that are difficult or impossible for humans, yet humans solve tasks that are impossible for machines. The point is that so far, the evolutionary paths of artificial and natural intelligence have been evidently divergent. The pragmatists' argument that for this reason we should limit ourselves to the exploitation of machines where they are effective, and only there, is practical and reasonable if the directive aims at the present, but it is dangerous if it implicitly implies that we must give up work that aims at the future and whose crowning achievement will be the automation of human creativity. Because no natural law prohibits it; only ignorance (a lack of knowledge) stands in the way. This task is of course beyond the scale and power of one generation, hence the psychologically understandable rush, followed by disillusion when the too-optimistic aspirations fail. But the task continues to challenge us, and that is why, sooner or later, it will be accomplished.

APPLIED CYBERNETICS:
AN EXAMPLE FROM SOCIOLOGY

There is no handbook yet for the pathology of socialist governing. People were made to believe that such a handbook would be "the socialism hammer."[1] But according to this reasoning, which ascribes the motivation for compiling the handbook to enemies of socialism, clinical pathology in medicine is inimical to health, and its handbook is a work of the enemies of healthy people. In reality, it is one thing to devise a plan for the systemic socialization of the means of production and entirely another to work out in detail the optimal dynamics of managing socialized goods. There is no deductive link between the two, just as the general theory of flight is not a blueprint for constructing an airplane. The fundamental assumptions of socialism, like the fundamental laws of the theory of flight, remain essentially unchanged, while models of social systems and airplanes must change according to the conditions of a civilization. The resistance to any rigorous study of the new system's inadequacies has led to further inadequacies, a problem that is addressed in this book, which deals with phenomena known from theory of regulation. When, in pathological processes, homeostatic systems, such as biological or societal organisms, deviate from an optimal trajectory, they do not simply stop functioning—they continue, but with regulatory aberrations, which include vicious circles and runaway perturbations. The vicious circle is feedback stabilized and supported by the pathological state. Runaway perturbation is the gradual spread of the "infectious" deviations from one subsystem to others that are linked with it. Consider a child with a vision problem: the child adopts an incorrect sitting posture when reading or writing to compensate, leaning to one side so that the better eye can see the blackboard; however, doing so only makes her myopia worse, and the altered tension in the back muscles stabilizes the pathological posture. Thus an aberration in one function brings about an aberration in another. Or take poor blood circulation: blood retention in one part of the vessel system causes problems that have no direct connection to circulation.

Systems without memory are insensitive to the frequency of regulatory interventions. An increase in the number of interventions always indicates an instability that makes the regulation of the system more difficult and more time-consuming—the control must be continuous, because a discrete regulation, based on a momentarily perceived deviation, causes oscillations in the trajectory. A car with a loose steering wheel is an example: the driver must make correction after correction, and the car's path resembles a sine wave.

Systems with memory behave differently: a system that "remembers" its past states becomes less responsive to regulatory interventions as their frequency increases. This results in an "inflation" of the corrections' interventional effectiveness. In a biological system, we say that it has become accustomed to a stimulus (e.g., a sleeping pill); in a societal system with high regulatory frequency, the high variability of laws and binding rules becomes functionally linked to the state of the system. This means that no matter how good the laws and rules are, an acceleration in their variability (too frequent corrections), having become a systemic variable, destabilizes the entire system (because the more variables a system has—and the fewer parameters—the more difficult it is to reach equilibrium). Therefore, preserving an imperfect system instead of replacing it with one that is unquestionably better is not as paradoxical as it may seem at first glance. Every law and established rule is de facto a systemic variable, because it can be *changed*. Whereas if a law or rule has been binding for a long time—longer, say, than a generation—then in the general consciousness it assumes the nature of a parameter, which, unlike a variable, is constant, given, and unchangeable. It then becomes a matter of course and will be included in the set of basic norms that regulate the community. A change in the law once in many years usually does no harm, but when the legislative activity leads to great normative variability, the societal reaction is a diminished trust in the legislature, because the convention-based and hence contractual character of the laws is then exposed.

The harm of great variability in established rules can be illustrated in the following, intentionally hyperbolized, example. In a certain country, let there be two parties that alternate in the governing and have opposing views on capital punishment. That arguments can be and are made on either side means it is impossible to say that either of these positions coincides completely with the society's sense of justice. Suppose, as a consequence of a series of crises, there are many changes in administration, and each time the opposition party comes into power, it immediately pushes through the parliament a change in capital punishment. If in two years the code changes eight times, then crimes committed in July and August may mean the death sentence, while the same crimes committed in September mean life in prison, and in October it is death sentence again, and so on. The sense of *injustice* caused by this alternation becomes pervasive in the society. As we see, this sense results not from weighing the merits of the arguments for or against the given form of punishment but solely from the frequency of changes in its application. Thus the variability in legislation above a certain frequency threshold can become a systemic variable that systematically destabilizes the equilibrium of a society (a system with memory), and perturbations in societal behavior correlate with diminishing faith in *any* normative activity. Upholding even laws that are obsolete may therefore not always be a manifestation of a societally harmful conservatism.

A slow variability in a law thus promotes respect toward it. Frequent change may cause another kind of perturbation when the time required for transmitting a quantum of regulatory information approaches the time in which its full effect can manifest. The latter time interval is usually long, as a law does not begin to act the moment it is promulgated but only when it is stabilized in the administrative structure *and* general consciousness. This happens rather slowly, according to the rules of "information osmosis," and an increased legislative variability only lengthens the process. *How* regulation is transmitted to a society may thus become a determinant of the behavior of the system *regardless* of the semantics (or contents) of what is transmitted. Consequently, theory of legislation might benefit from taking into account this cybernetic aspect of its functioning.

I now turn to how unplanned phenomena arise in regulation, in the form of its pathology, beginning with a simple model that can be simulated in a machine. When we link the inputs and outputs of several automata capable of learning conditioned reflexes, after a while one automaton becomes dominant over the rest. This process is random at first; the dominance is gained not by the automaton that sends impulses at the highest rate but by the one that raises its sensitivity threshold to incoming impulses the highest, thereby acquiring the autonomy—of apodicticity.[2] Thusly created dominance turns into a fiction when the automaton starts issuing directives that cannot be carried out. By the formal structure of connections, it continues to dominate, but the system exhibits a dynamic that is no longer consistent with that scheme. The probability that the dominator slides into a fictitious (apparent, formal) supremacy depends on the sensitivity of its feedback: the worse the dominant automaton is informed about the states of the subordinate subsystems, the higher the probability that it issues directives that are impossible to carry out. The entire system then enters a dynamic drift and creates operational rules that are a resultant of both the original structure and all the deviations that have risen. A mechanical system does not behave like this: the divergence between the planned and actual parameters leads to a steep acceleration of wear and eventually to a failure that destroys the system. But a system with feedback, even when damaged, usually finds other states of relative equilibrium, different from the optimal one, in which it can continue functioning though with varying degrees of informational and energetic degradation.

One of the systemic manifestations of regulatory pathology is the oscillations that are discussed in the *Dialogues*. Another may be the emergence in the system of functional aggregates that I would call "high-level informal groups." An "informal group" in sociology is a local set of people typically connected by personal ties of acquaintance, friendship, a common interest or aim, and so on. In such groups, nuclei of public opinion are formed, societally valuable ethical norms are internalized, and principal personal characteristics are shaped. Such groups constitute a practical, informal school of life for the members, along with their children and other relatives. They emerge spontaneously, in the absence of administrative-political mechanisms, and their structures can vary, as long as they remain relatively homogeneous. But I am defining a "high-level informal group" as a small community that has a dual structure: the real one, which factually represents the members' cooperation, and the formal one, which, existing only *de nomine*, often contradicts the fact of its existence.

As J. Szczepański has written, an authoritarian centralized government promotes the liquidation of informal groups on the elementary level;[3] I would add that at the same time it promotes their emergence on higher levels—of administration and the management of the society's common property. In general, attention focuses on informal groups classified as criminal: cliques that use societally privileged positions for private gain, either by balancing at the edge of the law or by accretion with an illicit economy, usually through corruption. But the societal pathology does not end there. When a society approaches a state of decisional and regulatory (mainly economic) paralysis, which can occur for many reasons, it automatically starts creating informal groups consisting of procurers of goods and public works. The heuristic rule that predicts the rise of such self-regulating groups says that the more inadequate the central government or the more it attempts to counteract the growing depression by issuing directives that ignore the actual situation, the higher the probability of the emergence—naturally, beyond the boundaries of the law—of what I would call manager groups. These arise mainly where they find favorable ecological conditions, as it were, where control is especially difficult, that is, not so much in individual factories, which are well defined and closed in both physical and functional space, as in border regions: between tentative investors, producers, and contractors of substantial works with a high processing potential. The phenomenon is typical in the spheres of urbanization, construction, and communications, and wherever cooperation of a large number of producers and providers is necessary to obtain a product. The primary purpose behind the creation and operation of informal groups on this level is, in a sense, noble in that they intend to facilitate publicly valuable works (housing projects, schools, factories, bridges, architectural complexes, etc.). Yet acting outside the bounds of administrative and regulatory channels, and thereby outside the law, they do not take into account the criteria for the optimal management of workforce, capital, raw materials, and machinery. Their operations may coincide with the overall economic plan with respect to its final goal but do not comply with the legally permissible structures of the division of labor and the official pragmatics of the process. They come into being as the praxeological quality of planned operations decreases because of the growing thicket of regulations, while at the same time the central government, accelerating the inflation of its regulatory commands, replaces the reality of the situation it actually caused with fiction, which is maintained for a variety of reasons. In a situation of such societal marasmus, potential investors and providers seek personal contacts to assemble, with confidence and mutual consent, the means and the workforce needed to realize the plan. People are selected on the basis of their managerial skills and on their willingness to work extralegally to further their personal ambitions. This is how, on higher levels of economic activities, the ancient system of barter in services and goods arises again, except that here the exchange and mutual promises of aid are based not on the pursuit of individual profit but on the wish to accomplish work that is difficult or impossible to accomplish under the binding rules of the government.

Where there is a permanent deficit in processing capacity or raw and other materials and a constantly overreaching production plan, managers build reserves that are carefully concealed from the central authority and its planners, knowing that they will have to act illicitly. The attempt to find cooperating parties through official channels is often doomed, because a contractor will refuse to provide the services he is formally required to provide, claiming, for example, an overloaded schedule. But if the requester has, or will have in the future, something to offer in return, he can count on cooperation under the condition of mutual trust, which is easy in the case of a personal acquaintance, because an extralegal contract cannot be put in writing. Thus cooperation among people in informal groups can often save a strained or endangered large investment, which, however, often means that another, unrelated plan will be put in danger.

The main gradients of the environment in which these activities are taking place should be briefly described. Central managers try to shape production-contracting subsets on the scale of the whole state in a way that at least roughly coincides with the material interests of the workers. But knowing that they will be unable to achieve full congruence between the vectors of public and private benefit (the state and the workforce), they intentionally employ political and patriotic motivations to amortize the divergence between the two. This kind of planning leads to ad hoc compromises in a space bounded by two extremes: one can be summarized by the slogan "Sacrifice one generation for the good of the next"; the other insists on the absolute convergence of the societal and the personal interests. It is a fundamental psychological error to think either of these two extremes in governing mass behavior is optimally effective. An *objective* congruence of the public and the private, determined by statistics and economic equilibrium, is not enough; if not accompanied by people's subjective conviction, even reaching the ideal state will not bring the maximum good—either to the society or individual citizens. Simply put: in order to work intensively and conscientiously people must genuinely believe that they are working for their own good and at the same time for the good of the society. The notion of "good" here can mean many things. Does it include the satisfaction from work as such, regardless of the pay? Apparently yes, but not always. It can also happen that work in the society's interest is easy and pays well and yet is avoided. Domestic service is an example: despite the privileges accorded to this kind of job, the high pay, and the fact that it relieves creative individuals—scientists, managers—from the mundane demands of living and enables them to do their socially valuable work, the status of such servants in the social hierarchy is low. An objective situation and society's views of it may vastly differ, which is precisely why the thesis of the congruence between the vectors of personal and societal interests may not work in practice. At the other extreme, "Sacrifice one generation for the good of the next," workers will respond positively only if they can see the effects of their sacrifice realized. Nothing crushes morale more than making people do work that is wasted in front of their eyes and thereby serves no one, either today or in the future.

When the central authority loses touch with reality because it has unknowingly initiated pathological governing processes, creating regulatory vicious circles and subsequent waves of perturbations, the entire economic organism of the society gradually shifts into trajectories of transformations that no one intended. A superficial analysis might suggest that the rise of informal groups in the economy is a positive phenomenon, a self-initiated form of addressing the growing problems. In this view, the only alternative to these groups is purely random behavior: when the economic plan exceeds the limit of the system's physical capabilities and therefore cannot be met by any means, some parts of it can still be realized, and which of those will be is decided either at random or through the silent consensus of managers "who happen to know one another." But this reasoning is wrong, and the dichotomy is false. Random realization of a part of a plan never takes place, because managers are not logically programmed automata but people. Each person first tries to act within the stream of legally permitted procedures; when their actions encounter resistance—the infamous "objective difficulties"—the still nominal collaborators become de facto competitors, as in the free market, except for one fundamental difference: neither the competition nor the collisions between individual managers' actions were foreseen by the central plan. The main criterion for realization of the plan then becomes something that had not been considered: personal contacts and relationships. The structure of a system overloaded with plans therefore privileges the clever, people who possess skills about which the socioeconomic theory of the system is silent; this is how the rule of creating personal relations *pro bono publico* self-selects. But why is this development dangerous? This type of economic management brings about the following consequences: (1) the normalization of extralegal solutions that evade or break the laws, as at least part of the decisions made fall outside the law; (2) a devaluation of general economic criteria, as the local managers care about their own problems and not about the interests of the whole country; and (3) the creation of a climate of demoralization, because evading the law, supposedly begun out of necessity, can easily become a matter of habit and pervade every aspect of life.

Subjective intentions are phenomena at the microsociological level, but objective consequences of realizing them can give rise to macrosociological effects that the actors did not foresee. New objects are built—bridges, factories, dams, and so on—thanks to informally made contracts. But additional processes are activated that are hidden from the central planner. Manager X agrees to build a new large project for city Y, even though he has no formal obligation to do so; he could plead a contractual overload or a mountain of current regulations to justify his inaction. But he accepts the job because he knows that it is indirectly to his—or to his group's—advantage. City Y is one link in an informal chain that connects members of the administration with the political system, so manager X is basing his decision not only on economic calculation. The chain of connections is reminiscent of the fable about the little rooster lying on the ground gasping because he choked on a grain. The little hen, trying to get the rooster some water, must beg many "parties" for various services and things, and after a great many of these red-tape steps she gets the water and saves him.[4] By doing the project for city Y, outside of his "portfolio," manager X secures for himself the gratitude of people who have connections with a group that can facilitate for him the acquisition of imported equipment that he needs to meet a plan that is in the core of his "portfolio." As in the cited fable, these links that influence managers' decisions are often longer than one or two steps and principially are not subject to optimization analysis.

In some situations, breaking the rules is a "lesser evil." Building a bridge requires the opening of dikes, but the water bureau takes time to give permission, wanting to avoid responsibility, so the bridge builder opens the dikes without permission, and there is a flood; then only connections and acquaintances can save him from serious trouble, first by providing an immediate supply of people and equipment on the site and later by sweeping the matter under the rug. In any case, had the builder not taken the risk, the bridge would not have been finished in time, if at all. This is the kind of decision-making that our manager faces.

Informal groups come together precisely in situations like this, operating according to an unwritten code for the exchange of services and promises, breaking various barriers put up by institutions, sometimes presenting what has been done as a fait accompli and pleading the gravitas of the public interest. These conditions and this climate are even reflected in the differential usage of personal pronouns, when the government is increasingly referred to as "they" and the work done in collaboration with it is less and less often labeled with "we." A patriotically educated first-grader might condemn this change, but a sociologist armed with observational and experimental material will distill its causes from the complex sociopolitical dynamics. The phenomenon is objective and its basis is not political (in the ideological sense): the informal group members use "we" when they speak about themselves and what they are doing but switch to "they" for the government because the government hinders their work. The third-person pronoun is for the dominant automaton that issues orders that physically cannot be executed; the first-person plural is for the subordinate automata that cannot influence the situation. This is an objective consequence of pathology in governing—since the process can be simulated in a digital machine, which surely does not suffer from any ideological conflicts or surges of antipatriotic sentiment.

This context also sheds light on the cause of what some time ago was summed up in the slogan "Poland is a loose federation of voivodeship committees."[5] As informal groups stabilized and spread, local party units shifted from political work to administrative positions, whereby they merged with the local system of economic managers, and the intravoivodeship congruence between the party structures and the hierarchical managerial structure facilitated this transition. Because the tactics of individual informal groups varied—as they had to, the process being spontaneous, with no plan or central control—the kind of cooperation between managers and politicians in the voivodeships varied. The administrative units started to differentiate—in growth dynamics, pragmatic efficiency, and fractional participation in the management of spending and effectiveness—whereby they acquired a spontaneous partial autonomy, as it were, of which the central government was aware but could not counteract because it was a resultant of mass processes of the growing pathology of governing. The voivodeships competed with one another for goods, capital, and the means of production, which led to systemic perturbations and uneven growth with positive feedback characteristics: those that had locally received still more from the central distributor as a reward for managing the received goods better. Had the relative autonomy of the voivodeships been granted systematically, with appropriate limits defined by law, the change would have been of benefit to the nation, and the differences in local economic conditions would have been addressed. But this was spontaneous drift, based on personal connections instead of governmental oversight, and thus was inherently variable and unable to guarantee that a work style once established would continue; uncertainty concerning the personnel in key positions, where some people's contacts were put above the established law, was an additional factor in systemic destabilization. Here we see how important law is in governing.

Functional equivalents of the informal group described above appear everywhere when there is a crisis, such as an environmental disaster that affects many lives. When there is a mining accident, a fire in an oil-storage facility, a famine, or a hurricane, rescue operations generally override the restrictions of laws and regulations, and the groups bringing aid do not worry about efficiency or the costs of what they are doing. But this break in the flow of regular activity is permitted only in unusual circumstances that cannot be foreseen or planned for, where the decision-making effectiveness of the government fails. Except that the growing unmet needs of the receptive economy create an environment in which informal groups operate in disaster mode for years, so eventually the other mode of action—in accordance with the existing law—becomes pure fantasy. Note that it is in fact the central government that preserves and protects this fiction, because it has no other choice. The press, which is under strict governmental control, sometimes informs the public about some marginal effects of these processes, for example that people can buy ham or children's shoes only if they know someone, or if they have "contacts," but this is like complaining about the low quality of roses when behind them a forest is on fire.[6] Members of the informal groups also preserve the fiction that the government is in control, because their power resides paradoxically in their pragmatic illegality. If they and their operations were uncovered and became public, the discrepancy between the reality and the fiction would precipitate a crisis that would end with the destruction of the fiction and the realization that structural reforms were necessary.

Because the scale of the reforms must be directly proportional to the duration of the systematic neglect, duration only lowers the government's decision-making effectiveness: the longer the dominator procrastinates in making the right decision, the more difficult it becomes to eventually make it. Something similar exists in biological pathology: the longer a disease remains untreated, the more difficult it is to bring the patient back to health; the longer the symptoms are masked or falsely interpreted—or even treated, instead of the cause—the worse the outcome.

We must also realize that the existence of informal groups is an open secret on the local level. Yet even the hordes of controllers sent from the central authority and issuing volumes of new regulations know that a sudden breakup of all informal managerial groups would, ending the economic lawlessness, also end the economy itself. Tearing off an ailing leg is not the way to cure a limp. The government knows that insisting on all the regulations would precipitate a complete paralysis of production and construction. Unable to disband the informal groups or even just expose them, which is impossible without sweeping reforms, the government falls into a state of permanent vacillation that turns the trajectory of its operations into a looped sinusoid. Such a situation favors the emergence of the pragmatic *lex ad hominem* rule: not everyone who breaks the law is brought to justice.

"The selection of coworkers based on connections" comes from ineffective government and chaotic clashes among hierarchical levels, but also from purely technological factors. Many final and intermediate products are supplied in violation of the rules and quality standards, and the receiving manager must either accept a flawed product or do without, and either insist on better goods and run the risk of retaliation from the beleaguered producer or put up with garbage.

In cooperating systems with many links, the following law applies: if just one link allows an inferior product to pass, the probability of low quality in the end product and in all its derivatives steeply increases, because of positive feedback. Defective car tires wear unevenly and cause not only consumer complaints but also more frequent replacement, under warranty, of the suspension systems, so the tire maker's technological error ricocheted to the maker of the car. Given the chronic lack of parts on the market, many cars keep running with damaged suspension systems and with tires that do not hold the road, so the rate of car accidents goes up, and nothing can prevent it in the given circumstances. The decrease in product quality also leads to a decline in sales and less profit, and thus hinders technological innovation, which is always expensive.

But still there are factories whose products are better or altogether respectable. The managers have a choice, yet they do not always choose a lesser economic evil, because their calculations are not determined only by profit and loss. They are subject to various levels of regulation, supervision, and quality control; their decision-making is marked by conflict between local factors and those coming from the center; and, although their function is administrative, they are evaluated politically and personally. These complications together make more difficult the scientific organization of labor, the prudent use of resources, and the solution optimization, about which we read many wonderful, exalted words in the press. What prompts manager Z to take on project V does not have to result from a rational evaluation of effectiveness and disponibility. Often the decision is a resultant of motives in which details that have very low importance or even no meritorious importance at all—such as who is on the board in institution X and who in commission Y—assume the critical role.

The organizational principle of the system is supposed to be universal planning, but the cumulative result of the described events is makeshift, patchwork, things done by whim, work surges or stoppages, and a demoralization that descends the social ladder. The government is torn between its lofty slogans and clear signs of functional collapse. Curious oscillating extremes appearing in the press mirror this split, as there are no sober analyses, evaluations, or criticisms, only unconditional praise or frustrated, desperate laments. Additional phenomena that arise include ad hoc theories and *ad usum delfini* practices,[7] journalism indulging in self-flagellation, and criticism of the national character (Polish laziness and anarchy, etc.). Such behavior is unreasonable; it is like accusing passengers forced to travel a long journey in a dirty, unheated, crowded, and poorly lit train of being grubby, argumentative, with poor manners and repulsive looks; in such conditions only a saint could retain good humor, elegance, a neat appearance, and kindness.

Neither official decision-makers nor members of informal groups realize that this method of work, which has been establishing itself by carving its streambed of illegal operations for years, is leading the nation toward a civilizational-technological crisis. Increasingly the "work-surge" style, the reliance on personal contacts to select coworkers, and making objective profitability a lesser priority are out of step with developmental trends in the rest of the world: improved coordination in production, less tolerance for inferior quality, and the persistent effort to innovate so as not to fall behind others.

Maintaining the fiction under these circumstances is a path to self-destruction. Unfortunately there is a characterologic type of people who flourish precisely in such informal workgroups, where they indulge their ambitions and, although aware of the risks, they receive satisfaction from manifesting their pugnacity. They behave more like gamblers or guerilla fighters than organizers and economists. They prefer to work outside officially established structures and use their intuition. They care not about actual costs, but only about the costs that cannot be hidden from the central supervisor. They know that the more urgent or difficult the matter and the higher the degree of involvement of the highest levels of government, the greater the danger to them, but also the chance that if they succeed, their evasions of the official pragmatics or even transgressions in competency and economy will be overlooked. Obviously such people bring pragmatic benefits but also long-run damages whose extent defies estimation. Such behavior first demoralizes the manager himself, since breaking the law and ignoring the rules, even with good intentions, makes him habitually think that rules only hinder precious initiative, which sends him on the path of a downward spiral of undermining all norms. The deformation of economic decision-making structures initiated by the informal groups creates new dynamic systems of which no one is fully aware and therefore no one controls. Once these spontaneously formed dynamic relationships have stabilized, any attempt at socio-economic amelioration will meet with resistance from the groups, because they will consider any change a threat to what they have accomplished. They also know that their way of operating can never be transformed into any officially permitted or legally supported practice—by any reform. Moreover, the government cannot really reform the existing situation because it sees only structures that de facto no longer exist, as the above-mentioned activities have eroded them away.

Many disponents of public property who work in the manner described above have a subjective feeling of being unappreciated or even wronged, since filling the role of a high-risk player, in the long run, can give satisfaction only to exceptional individuals. Their opponents will gradually include both the superior and subordinate institutions because informal groups require a certain level of secrecy, since they are constantly on the edge of illegality. These people have elevated opinions of themselves and believe that the country endures because of them and that their intensive efforts are not for their personal benefit (which may often be subjectively true but does not change the fact that their behavior is harmful from the point of view of societal education). This overly focused cult of the pragmatic, strictly operational proficiency carries such high economic costs that the economy begins to crack under its cumulative burden. There are no ideals worth following, just random grouping and ungrouping in response to changing political and personal fortunes. The moral subtext of these tactics is that no good for the country can be done in a lawful way and any trick, any lie, is justified to achieve the desired result, starting with supplying false information to one's superior.

But the path to the higher societal organization does not lead through cronyism; services should be provided according to the letter of a law that reflects the overall societal interest. It is wrong when without knowing someone you cannot get ham or children's shoes in a store, and it is wrong when without the right contacts you cannot build a power plant or a bridge. As a side-effect, the personal benefit of prestige and influence gained from occupying a high position then facilitates obtaining services that otherwise take a long time. Yet even though the boss's subordinates are busy arranging personal favors for him and attractive goods circulate only in closed circles, the economic damage thus caused is marginal on the largest scale and may not hurt the country as much as the injustice to those who are not well connected and must stand in lines that get longer and longer. The social and moral harm is worse than the harm of chaos and waste in economic management. It is then understandable and proper, but also rather unimportant, that the central authority tries to root out Lucullian pleasures at the government's expense because even the most unbridled gluttony (parties, banquets) of the managers will not bankrupt an economically solid country.

The kind of economic management practiced by the informal groups is similar to the spontaneous expansion at the beginning of capitalism, which was marked by zeal to destroy the feudal structure that gave birth to it, and which was born not in a country with a strong government and tradition of law but at the periphery, where the law system was poorly defined or unenforced, where the survivors were the powerful, skillful, sly, and unscrupulous. These personal characteristics then become assets mainly when such people gather, often in secret, to pursue common goals, among which is the destruction of their competitors. The economic management practiced by informal groups also resembles the expansive dynamics of natural evolution, except that the habitat of the manager is not nature but a civilizational landscape that is being mindlessly devastated, mindlessly because the mismanagement is not caused by personal greed or any motive that might be identified as a kind of social Darwinism. The squandering is caused by operations of an informal group that is obliged to perform specific creative work but at the same time carefully conceal all illegal methods and acts that assist their work. Because for the group in that situation the existing law becomes just another obstacle to overcome, and waste is a lesser evil—not by choice but out of practical necessity, since the simplest and most economical path may be unavailable to the group. The managers' field of vision narrows down during the enterprise. Economic and ethical criteria both disappear from their horizon, and what counts is immediate success; the price is unimportant as long as it can be concealed. As an indirect consequence, the chaos caused by the last-minute work surges acquires an autonomous value, as it aids in camouflaging the endeavor's characteristic illegality.

Years of such activities result in the formation of pragmatic rules of an exclusively extensive economy that pays no regard to future consequences, especially global ones. For the central authority, which keeps an eye on the interest of the nation, the critical problem—the modernization of production to make the nation's products competitive in the world market (for economic autarky is impossible for midcaliber countries)—reduces to a dilemma that in reality is false. It assumes that production can be divided into two separate enclaves, one for domestic consumption and one for export (the quality of the exported goods being higher). This division is detrimental to a system's stability. It is effective at a low level of industrialization, when most of the export is raw materials and intermediate products, but has negative consequences when the export goods are modern equipment, which is usually made through the collaboration of many parties, because such a system is more vulnerable to perturbations whose causes may be global. A drop in the market after a boom is always possible; the pattern of demand for energy, raw materials, and intermediates is subject to oscillations, and, as we know, only a regulator with sufficient internal diversity can counteract them effectively. If someone prefers a narrow sector of export production, he is making himself dependent on the boom's stability, like the farmer who has only one crop. When the average industrial production is high in all sectors, every sector serves the nation as a reserve that can be tapped to maintain the balance of trade. An export gap can be closed with almost any kind of product originally intended for domestic consumption—radio sets, cars, refrigerators, machine tools, and so on. But if those products are poor in quality, they will have no export value and the economic stability of the entire system decreases due to increased sensitivity to external perturbations.

A dynamic equivalent of a fully industrialized country, which possesses a rich repertoire of exchange possibilities with the rest of the world, is natural biocenosis, because the country's equilibrium is a result of the mutual cooperation of an enormous number of simple homeostatic mechanisms that in nature are the individual species and biotopes. In contrast, a country that divides its production power into domestic and export is similar to artificial biocenosis, like a farm with a monoculture of a single crop, and as we know, the homeostasis of such a system is difficult to maintain, since it is highly susceptible to a large number of harmful perturbations, which often cause oscillations in the monoculture—the amplitude alternation in fat and lean years. Taking an ecologically poor system out of such oscillations once they started is extremely difficult.

One more thing should be said about an economy that is developed under the aegis of rigid planning but is de facto lawless: the more you have, the more you can squander.

The reluctance or lack of knowledge about how to change radically for the better can lead a government to throw the responsibility for the increasing hardships on the shoulders of the masses, not out of malice but due to the fact that this is the easiest course of action. But the masses coaxed into systematic ascesis begin to interpret the situation incorrectly—as a sign of ill will or outright malice, which only deepens the schism between the governing and the governed, both sides being equally in the dark about the real state of things. When a pipe between a pump and a tap leaks, two interventions can be made: increase the pressure or shut the valve. An increase in pressure will mean more loss of water; shutting the valve will mean no water for the thirsty. The right solution, obviously, is to inspect and replace the pipe.

Finally, I observe that for a spontaneous-surge economy with public aims but the means of attaining them kept secret, which is characterized by antagonism between the central planners and the subordinate managers, a statistical-analytical approach will fail, because statistics can depict what happens in a closed economic unit like a factory or mine but not what happens at a unit's multiple contact sites. The inadequacy of a product that proudly fills a statistical table becomes evident not in the factory but elsewhere. When bad tires cause accelerated wear in and damage to a suspension system, another manufacturer has to pay. The more complex and chainlike a system of relations is among manufacturers, the more difficult it is for statistics to find the defective link. Hidden dynamics resist the methods of classical statistics, especially when the data from the highest level are belied by the data at the lowest.

Change is certainly possible, but it is a lengthy and thankless task. The order of actions to take depends on a system's past. The first step is to recognize the situation, that is, to assemble a map of dynamic events in their actual structures, which differ from the formally supplied blueprint. The second step is to change the structure in order to secure a correspondence, in terms of economic correlates, between the developmental plans and the delegated competencies with the aim of creating operational reserves because the mere existence of accumulated reserves cuts the ground from under the feet of the informal groups, that is, it prevents their formation behind the scene—objectively, not only by established laws, because the groups operate outside the law anyway. Only the third step represents the psychosocial transformation of people's opinions, and it must follow the first two, not come before—because an inflation of the appeals to the conscience of the governed is unfortunately already a parametric variable of the system. The inclination to keep issuing precisely such appeals is psychologically understandable, but their effect will be a paradoxical response, opposite to what is expected. Due to the obstinacy and sheer volume of the slogans heard for years, people close their ears in a commonsense act of self-defense, unable to believe the words that everyday reality negates. The appeals then activate the conditional reflex of suspicion that everything is just starting over again. Hence strictly objective calculations based on knowledge of the laws of macro- and microsociology of governing and motivation dictate restraint here. Only a systematic segmentation of societal transformations, in the correct sequence, can turn a Sisyphean task into a project that equally satisfies today's needs and tomorrow's ambitions. If this simple truth is not understood, there can be no remedy.

SUPPLEMENT 2: ADDITIONAL ESSAYS

THE ETHICS OF TECHNOLOGY
AND THE TECHNOLOGY OF ETHICS

Introduction

1. The acts of a lion killing a lioness, a rabbit doe eating her kittens, or a praying mantis female eating the male after copulation usually are not considered unethical, because we do not put animal behavior in such a category. Yet we may find a difference between the first two examples and the third: what for the praying mantis is a species-specific behavioral stereotype is a departure from a stereotype for the rabbit and the lion. The difference is based on the idea that animals—in view of the goal of species preservation, which is an evolutionary given—cannot behave in a manner that leads to their species' extinction. In the sense that it is teleologically conditioned, the stereotypical animal behavior is rational—it would stop being rational, for example, only if the female mantis ate the male *before* copulation.

2. Human ethics contains a similar rational nucleus, but it cannot be considered a stereotypical species behavior simply because no such unified stereotype exists. Ethics appears to be a consequence of the emergence of language, which enables us to compare present "model" situations with those that have taken place or are anticipated. If the model is "appropriate" or "inappropriate" (distinct from the criterion of true or false), it is possible to evaluate it axiologically. And if the model represents an interpersonal situation and the comparison is made to determine its consistency with behavioral directives (stabilized by culture), then it acquires an ethical character.

3. Which situations are subject to ethical evaluation is decided by culture. For example, personal unemployment is in some cultures ethically neutral, while in others—especially in the industrial ones—it is blameworthy. For example, ruling cultural patterns of behavior may

require constant activity, what's more, an activity of a specified type: in some cultures, a *sensu stricto* creative work is esteemed, while "doing business" is looked down upon.

4. By "ethics of technology" I mean, in this essay, the effects of technological development on the ethical behavior of individuals in society. When speaking of "ethical norms," I mean those that can be "distilled" and reconstructed from empirical studies of individual behavioral stereotypes in ethical situations, not the norms that people stereotypically declare in their oral statements. A societally declared ethics is not necessarily identical to the one that the society practices. Such divergence between a theoretical formula and an actual stereotype used in practice occurs in all societies, and if they are stratified (by class, occupation, etc.), the divergence, which is to some extent adaptive, can create various groups, classes, or professional ethics. The difference between the ideal and actual behavior is certainly an important parameter of any culture, but I am not addressing that topic. Speaking of the effects of technology on ethics, I limit myself to the changes occurring in "ethical behavior" without giving much attention to how they are viewed by education, propaganda, or religion.

5. Readers might get the impression that the effects on which I focus work like this: ethical norms A and B allow for the development of technology X, but after some time it turns out that the technology has pushed norm B out of the system and replaced it with the new norm C. The system (A, C), different from the original (A, B), can be called an ethics transformed by an instrumental effect or, in short, "the ethics of technology X." But, except in special situations, technological processes do not affect ethical phenomena in this way. A change in ethics after a societal transformation caused by technoevolution is marked by adaptability that is extemporaneous, and therefore ethics is a behavioral program that undergoes transformations on a level that is different from that on which ethics actually operates. Here is an analogy from the organic world: a change in ethics corresponds to a transformation that creates a new species: the factor that induced the variability does have a connection with the creation of the species, but adaptation is not a simple result of inheriting acquired traits. Like genotypes in biogeocenosis, people in a society have at their disposal an

enormous excess of diversity in their response to situations, a diversity that can exert a regulatory effect if a need arises. In a genotype, variety exists thanks to the reservoir of recessive genes, which is continually being enriched by mutations; in *homo socialis*, the variety is due to a behavioral plasticity (the "reactivity potential"). In an evolving culture, technology and ethics appear to be *dependent* random variables, so we need to study changes in both and at the same time heuristically accept their stochastic nature. Such study is difficult to carry out, because in a complex system like society causal chains often fork, and we end up not with a chain but with a network. In that case, selecting specific links and connecting them in a single chain will always be somewhat arbitrary. Therefore, instead of seeking technological causes and linking them to ethical consequences, we should look for correlations. To my knowledge, no one has done this in a rigorous and well-documented way. For example, there may be a connection between the ethics gravitating toward nihilism in some youth groups and the "technological explosion" in this century, but this hypothesis cannot be subjected to "falsifiability" tests.

6. In the second part of this essay I deal with the "technology of ethics" understood in two ways: a search for technological tools for modeling ethical phenomena in a research program to study cultural/societal phenomena in a substrate that by itself is neither "societal" nor "human," and an attempt to harness instrumentalities to serve ethical directives.

I

1. Since technology rearranges the environment in ways that make it conducive to human existence, it is an extension of natural homeostasis, since there is no principial difference between the five senses and sensors of research instruments or between muscles and engines. The senses and sensors both receive useful information from the environment; muscles and engines, guided by that information, both enable energy independence from the environment. But technology, once set in motion with the aim of "satisfying needs," increasingly tends to facilitate all kinds of "satisfaction" that we can imagine. From a

strictly instrumental point of view, there is no significant difference between satisfying the hunger for food and the hunger for sex, since both are biological. Technology, which entered the area of interpersonal relationships a long time ago, is now, in the next step, penetrating into more private spheres of our existence—with ambiguous results. We are finding out again that the sequence, in which we conquer specific segments of nature, including our own bodies, and which nobody rationally planned, can hide antinomic traps.

Technology offers choices where only fatalism existed in the past. In the not too distant future, we will probably have the option to decide the sex of an unborn child. The equilibrium ratio of sexes in humans has been regulated, as in any "undomesticated" species, by probabilistic chromosomal automatisms. But if parental decisions about the desired sex of the child depart from the ratio determined by those automatisms, for example, due to a cultural preference for one sex over another, the existing equilibrium will be disturbed, and we will need to take steps to restore it. This is one example of a general phenomenon: when parameters that have been kept in a homeostatically beneficial interval by "natural" regulative feedback (i.e., without human interference) are removed from the control of those automatisms by a new technology, "artificial" actions may be required to keep the parameters in that interval. An "artificial" action may be one that limits the freedom of an individual that was just recently expanded by the new technology. In this situation, the peremptory simplicity of the original, ethically neutral statement "It is not possible" (i.e., to predetermine a child's sex) is replaced by the directive "It is not allowed" (albeit technically possible—unless, say, the quota for a choice has been reached).

What, then, about the possibility of selecting other physical and mental characteristics for an unborn child, which biologists currently anticipate (i.a., Rostand has written on this topic)?[1] It would be extraordinarily difficult to satisfy the wishes of the parents while at the same time keeping in mind the welfare of society—a society consisting solely of geniuses would probably not function in any kind of equilibrium. But changes that would have to happen in specifically human values should be considered even more significant. Suppose, for example, it becomes known that the exceptional talent of Mr. X is not the result of a "chromosomal accident," or, one might say, "winning the jackpot in the inheritance lottery," but instead is due to the permission, obtained by Mr. X's parents from the appropriate authority, to add this talent to their child's genotype. Objectively there is no principial difference between the natural genius of today and the engineered genius of the future, because in both cases, the talent's cause is external to the individual. A great composer is a great composer whether his genes assembled themselves "on their own" or were assembled by the genetic engineer (with proper administrative permission). Yet it seems that an instrumental intervention that would privilege some people over others creates in the public consciousness a feeling of injustice, because not everyone would obtain what almost everyone lacks, that is, the "talent gene." The novel of a "synthesized" author might still draw applause, but many readers might feel a rather general aversion for the author's persona. Once "genetic composition" technology is initiated, changes in the societal system of values that are called autonomous must inevitably follow. But let us leave the worries about that to the future.

2. Two-thirds of humanity, two billion people, are chronically underfed, and out of those about forty million starve to death each year. At the same time, elsewhere, an excess of food shipped to market makes it necessary for crates of it to be destroyed. Yet it is not correct to think that the poor are miserable and the rich are happy. In reality neither are happy, although the effects of surfeit and want have little in common. We tend to take want seriously but disregard, or consider humorous, the dangers of surfeit. It is understandable: our species evolved in conditions of a continual struggle to meet elementary needs, like all "undomesticated" forms of life in nature. The situation where hunger and all desires can be satisfied too easily is a true novelty in our history, and until recently it was held to be a good thing. But we are learning now that surfeit can be detrimental to the values that constitute the motivational skeleton of human behavior. The negative impact of a technological satisfaction of needs is sometimes obvious. For example, a few micrograms of LSD (lysergic acid diethylamide) induce a state of subjective bliss, almost a mystical fulfillment, unlike anything else. Human beings are always looking into the future, and

the meaning of life is shaped by expectations, hopes, and desires. LSD, by removing all personal anticipations, amplifies the existential experience of the present so much that everything else seems meaningless, as if someone reached the peak of a mountain that had never been climbed before. A comparison with the effects of LSD on insects will be instructive. A spider under the influence of LSD keeps weaving its web,[2] but the web is much more geometrical than normal: the drug cuts off external stimuli but does not interfere with the instinctive behavior that has been established once and for all by genetic programming that can now manifest itself in its "purest" form. A person under the influence of LSD, however, loses all ability to act in the real world, because his motivational mechanisms, which are not inborn but created by cultural imprinting, are much easier to dissolve. A state like this is detrimental because it leads to breaking all ties with other people.

Because a society consisting of individuals under the influence of LSD could not function, the drug has become a threat for society, especially in the United States, where millions of young people use it; it was declared a narcotic (even though technically it is not), and its distribution was made illegal. It is used, experimentally, to ease the suffering of the terminally ill, and it works well: people become indifferent to death though aware that it is imminent.[3]

More or less at the same time, oral contraceptives were introduced in the United States. The widespread use of these pills has not shown any adverse effects on the body, and prima facie it is not clear why anyone would object to something that separates procreation from the pleasure that evolution has attached to it. Considering the current growth in world population, the pill has come just in time. Previous contraceptives were lacking in esthetics and reliability, whereas the pill can be taken like a vitamin tablet, and, what is more, it can be used *post coitum*, which is a great psychological advantage (the woman may not anticipate the possibility of sex). Men and women now have equal rights in the biological sense as well, as both can avoid the consequences of copulation.

But for both LSD and the pill the transformative effects of technology have negative aspects. The chemically guaranteed barrenness of copulation (not omitting other factors, about which more later) may lead to the weakening of the relationship between the two sexes, just as LSD severs a person from other people. The problem is not so much the value of the "experience of the absolute" itself or that it was chemically induced as that this technological intrusion achieves full satisfaction through purely local action. Because local actions may have distinctly nonlocal consequences. The use of pesticides to kill certain insects ended up shaking the entire ecological pyramid of the species in a given area. Insecticides cause an imbalance in the material system of ecological hierarchy, and chemicals that quench desires and motivation can cause an imbalance in a society's axiological system. Making the sexual act "safe" by removing its normal consequences adds to the ease and casualness of sexual relations that has already been taking place in our culture. The historical values regarding sex are due not to inborn mechanisms but to an internalization of particular attitudes that have ethical dimensions, have been approbated by society, and therefore deemed valuable. That the imponderabilia of eroticism are conditioned culturally, just like the complex and often painful initiation practices in primitive societies, does not mean that they can be dismissed as irrational and dispensed with. In fact, all culturally relativized values are "unnecessary"—but only in the sense that in different cultures, different values serve the same purpose. The obstacles that a society places on the path to an individual's maturity (group, family, professional, sexual obstacles) are not just "superfluous complications." By removing them, we are at the same time destroying attitudes that motivate, and usually without offering anything in return. Technology is more effective in destroying autonomous values than in creating them. Forcing technological "improvements" can therefore initiate an "axiological implosion," a collapse of an entire value system. It may lead to a life that is effortless but not worth living.

I am not saying that the contraceptive pharmacology will destroy erotic love. There surely are cultures axiologically constituted such that the pill has no "value-killing" effect. But in ours, given the trend mentioned above, the pill is a factor that makes "loveless sexual situations" more probable. The statistical aspect of this phenomenon is essential, because it determines the developmental direction of ethical change. It is true that there have been women who were chaste only because of the fear of pregnancy and did not really respect the immanent values of eroticism, due to which the abandonment of sexual ethics will now surge in the highest level of externalization—in behavior. But this factor does not seem important to me. Ultimately it is mass behavior that decides the hierarchy of societal values, not the analysis of individual motives and attitudes. (I sidestep the questions of which is more important, what people themselves think that they are doing or what others, e.g., the psychoanalysts, think about the reasons for their actions; whether the "spontaneous" self-knowledge of the common people or the "professional" introspection that philosophers practice should be the starting point for such analyses, etc.)

3. Knowledge today is gained through long and arduous research. When an "information pill" supplies people with knowledge, such research will become superfluous. The technology of "free learning" does not yet exist but appears to be a possibility. The toil of learning, however, has a role beyond the acquiring of information capital. It trains people to overcome obstacles, cope with stress, and improve their character. Therefore an "information pill" could harm a person's mental development by providing erudition to a mind that is fundamentally unprepared to make full use of it. A new form of education would then be needed: How to benefit from the information just ingested. Or, what admittedly borders on the absurd, a direct intervention in the brain might arrange its processes into the same state that would be achieved through "regular" learning. Yet if it were possible to obtain a universally proficient mind through a sequence of instrumental (pharmacological or electrochemical) procedures performed on the brain, what values would remain in such a world to give life meaning? Creating shortcuts for all possible needs, desires, and wishes should not be the purpose of technology, because where everything is available in an instant, nothing has value. Value originates only from a hierarchy of goals and a gradient of difficulty that must be overcome to attain them.

Meanwhile technology is invading us on many new fronts, and it is not clear how our bodies should defend themselves, as the besieger appears to be the friendliest of allies. The philosopher Pangloss may not have been correct two hundred years ago,[4] but we are now approaching the best of all possible worlds at the speed of a cannonball, a perfect place where pharmacies will distribute knowledge without learning, mystical states without faith, and pleasure without scruples. In this modern version of a mercenary economy, convenience replaces value. It is hard to object to the introduction of contraceptives, since desperate situations demand desperate measures, but it is necessary to call a spade a spade: technology cannot replace the axiological backbone of civilization. In the modern world, customs and moral norms are unable to resist the pressure of technology; they can slow it down (as in the case of LSD), and even that only when the instrumental innovation's effects directly clash with the code of established laws. But technology, instead of launching a direct attack, takes roundabout paths that render societies and their laws practically helpless. And the damage, once done, cannot be reversed. When a technology becomes omnipresent, people grow accustomed to it and would consider its absence almost as an injustice. Changes in ethics occur gradually and without plan. I do not know if anyone has studied the socio-ethical aspects of the liberation of atomic energy, for example, the attempts to draw parallels between the genocidal practices of the Third Reich and the first use of the atomic bomb by equating the creators of the gas chamber with the creators of the bomb—because they all were scientists and engineers, weren't they? In the area of societal practice there is no prognostication, only randomness; no control but at best a "concerned" passivity; and in the place of knowledge there is an ignorance that is barely even aware of itself.

4. One might argue that these lampooning remarks on how technology degrades ethical norms should be accompanied, for the sake of completeness if not anything else, by an apologia of technology's positive influences. We know that advances in energy, transportation, production, and distribution of goods all promote global cooperation and are therefore not only morally commendable but also economically beneficial. Unfortunately, the many antagonisms of the modern world can put an end even to something that is materially beneficial for all. I devote more attention to the negative effects of technological development on societal value systems because they are more difficult to

recognize. The role of modern technology is particularly questionable in the Third World, where many cultures remain at relatively primitive stages of socioevolution. Traditional norms, incompatible with demographic changes taking place now and helpless in facing new situations, are vulnerable to quick erosion and collapse, which may easily create a normative vacuum, a disappearance of the old values without the appearance of new ones, since it is not possible to speed up artificially the development of ethical norms. Yes, children learn ethics as they do their mother tongue and later the natural sciences and mathematics, but these are very different kinds of learning: behavioral rules are adopted not by remembering information but through the continuous observation of social patterns. When a civilization evolves, climbing from one level to the next by its own efforts, the pace of ethical growth will be organically slow and harmonious. The sudden invasion of technology into a primitive culture may cause ethical havoc, because the adaptive mechanisms of customs and morality cannot keep up with the changes. But even a continuously developing civilization can—due to the technological acceleration—reach such a technoevolutionary speed that customs inculcated at an early age may not last a lifetime but become obsolete, and the next generation, raised by axiologically disoriented parents, seeks behavioral goals on its own, often with poor results. I do not know if the pace of the evolution of customs has already been overcome by the pace of technoevolution or if it is just going to happen. But the constant acceleration of instrumental progress makes this divergence, this loss of intracivilizational coherence, a very real possibility.

Nature preserves its equilibrium and continuously renews itself, using its own elements; its highly stationary character is the result of very long processes, spanning billions of years, of coadaptation of geological and biological factors, partly by transforming the former into the latter (this is how the biosphere arose as a homeostatic unity) and partly by adapting the latter to the regular inanimate fluctuations of the planet. Human beings, infinitesimally small in space and time, have always treated nature as an open, inexhaustible system. Yet, although the mass of all living human beings is a tiny fraction of the planetary mass, human technologies have transformed this open system into a closed one and made the stable equilibrium of the biosphere unstable, hence the emergence of new technologies whose only purpose is to mitigate the damage caused by previous technologies that directly serve the human biological and social needs. One can imagine that a demand will arise for a next wave of technologies whose purpose is metamaterial—that is, technologies to counteract the phenomenon of "instrumentalisms going out of control," which puts the front of the actualized causative possibilities of the civilization beyond the reach of any of its individual members, beyond the traditional axiological horizon, and beyond people's abilities to absorb and adapt, which are truly enormous but still exactly the same as those of the Mousterian and Aurignacian cultures[5]—because biologically we are equal to their members. I ignore here the epistemological and cultural-educational aspects of this acceleration, the problem that the swift obsolescence of knowledge, including professional skills, will make constant learning and retraining in many occupations a necessity. This is already happening in the natural sciences: progress has forced biologists to learn mathematics, economists to master information theory, and so on.

5. Empiricism should be subordinate to ethics in the sense that through discovering links that are imperceptible in everyday life, we begin to get an idea about the ethical weight of acts and decisions that were previously considered ethically neutral (if a physician reveals to a young couple that their offspring will very likely have a genetic defect and they go ahead and produce physically or mentally disabled children, they may be innocent in the eyes of the law, but their ignoring the physician is morally questionable). It is admittedly strange when ethical is not what your conscience prompts but what has been sanctioned by the biologist, the system theorist, the expert in decision-making and linear programming, and the cyberneticist working in the field of game theory. Granted, we do not face such problems every day, and traditional morality, especially in our daily contacts with other people, has not yet been completely lost in the forest of instrumental directives and facts. It is still possible to be a good person in the ordinary sense, but unfortunately the sensitivity of our moral compass is under a constant attack, since the technologically-enabled global news makes us witness dreadful things happening in thousands of places

and we can do nothing except shake our heads—and everyone knows that that is not enough. Belonging to the species *Homo sapiens* today can thus be regarded as being responsible for the species' overall fate but having infinitesimally small personal power to influence it. This disparity is a consequence of the many technologies that have linked each one of us in an excessively unidirectional way to the other three billion members of the human world.

6. To speak *in full* about the "technology of ethics" would mean to explore the theory of an ideal society (by analogy to an ideal gas), because ethics is a part of a culture, a subset of regulatory parameters, which can be defined by conventional and simplifying abstraction, of the system we call "humanity." Current knowledge does not provide sufficiently solid basis for such an endeavor, unless one remains within the limits of what can be empirically verified. So my discussion is no more than an introduction to this "technology"—the (formal) modeling of the phenomenon.

II

1. I see ethics as an unwritten part of the rules of the "game of society." Some of those rules undoubtedly have an instrumental character, but whether or not they also have an ethical flavor depends, among other things, on the totality of the culture. Since ethics resides in interpersonal relations, which social situations possess ethical elements and which system of valuation applies to them is clearly defined when viewed from within a given culture; but many situations classified as ethical, together with the classification criteria themselves, prove to be variable (though not without limits) when examined from another culture. Observers from outside of a given cultural circle will offer divergent evaluations of interpersonal relations in that circle, which necessarily means that they have a different cultural imprinting. To refrain from judging another culture from the viewpoint of one's own is possible only if the observed phenomena have no cultural significance but simply reflect an equilibrium behavior of elements in a highly complex material system. One might try to be as objective as possible even without resorting to such an extreme, physicalistic

atomization of the "human set" with the extreme objectivity of scientific analysis, but there must be limits to it that no one is willing to demarcate because no one really knows where they may lie: what in our behavior is "metacultural," and therefore free of any relativity, could only be revealed by experiments that cannot be conducted—for obvious reasons of an ethical nature. Comparing as many cultures as possible, however—those that have reached the same technological level and share some aspects of their development, such as similar ecological environments or anthropological ancestry, but also those that developed in very different conditions—shows great promise.

2. Anthropological research has shown many times that the biological differences between human races—relative to the cultures that the races create—are practically negligible. So if the compared cultures share the parameters in the areas of geography, climate, and technology, the comparison is supposed to reveal whether, in the absence of other significant factors, the structures and developmental trajectories of the two communities coincide, as one would expect. But as we know, this coincidence does not occur: in terms of customs, beliefs, and ethical and esthetic norms, primitive cultures (for they are the topic here) greatly differ from one another. Certain basic principles are certainly preserved in all, such as the principle of cooperation, which is, in a sense, both trivial and obvious, since a society that opposes all forms of internal cooperation cannot exist. The parallels that have been observed are limited precisely to those principles whose violation is, for purely biological reasons, impossible. A logical, rather than empirical, analysis suggests that the cooperation principle must have been the seed of cultural development; we might conclude then that cultural differences arose because societies took different paths to the same technological level (inventions and discoveries made in a different order, different ways of setting snares and traps, different ways of hunting and building shelters, etc.). But this is not the case. Cultures, even if they were "constructed" around the primary principle of cooperation, manifest rules of behavior that are definitely superfluous to all instrumental activities and therefore cannot be reduced to that single guiding principle or to the specifics of the methods used to manufacture tools, work the land, and so on. For unknown reasons, some

cultures are patrilinear and others matrilinear; some practice ethics that Western scientists call "Apollonian," others "Dionysian." We can make many comparisons like this, since about three thousand different primitive cultures have been cataloged. Each culture has its own "ideal" of what a human being should be, and the range of these ideals is surprisingly broad.

The question is whether a set of tools and the ways they are used can serve as an impetus for the development of a culture's activities that are, from the economic point of view, "superfluous." Such a set would act as a seed for crystallization of activities that can later become autonomous, for a growth above and beyond the satisfying of actual needs, that is indeed unnecessary from the rational-engineering standpoint but justified by the mentality of people at a low developmental level (open, say, to animism or magic). Importantly, a mix of the irrational element, with aims that are physically fictitious, and the rational, instrumentally teleological element is found in many primitive cultures, but it does not explain why certain societies practice "Spartan" ethics (even its most extreme forms), while others—equally developed in terms of the intellect and technology—have created ethics that to us seem liberal, approaching the Western ideals of humanism, where the leading directives are to be kind and gentle to all. The very fact that this question is posed indicates that nothing like "immutable human nature" exists in the real world and people are neither "immanently good" nor "immanently bad" but only how conditions make them. But I ask again, where do the differences between societies, often shockingly wide, come from? A series of experimental studies, by the way, conducted outside the boundaries of anthropology—in theoretical biology, in the form of the computer modeling of bioevolutionary phenomena—suggest an answer.

3. Modeling evolutionary processes within the Markovian framework is powerful. It usually uses a relatively simple form of A. A. Markov's probabilistic (stochastic) process with random dependent variables that is called the homogeneous Markov chain (but as we will see shortly, the genesis of a society and culture cannot be modeled so simply). Processes are called Markovian if the prediction of a future state is determined solely by the knowledge of the current state

and no information about all the previous states is necessary. The same process can be Markovian when described one way and non-Markovian when described another way. If we are dealing with the development of a population, its purely phenotypic description is not Markovian, because it does not include the information about recessive traits; when described at the genetic level, the process is Markovian. A non-Markovian description usually omits some parameters that are essential for the system's behavior. For example, predicting a person's behavior on the basis of knowledge of his past is non-Markovian, but a forecast of his behavior can be made on the basis of a detailed study of his brain, with all its neuronal preferences in stimulus transmission. The latter description is Markovian and would not contain the word "memory," because as Ashby noted, "memory" is just shorthand for parameters that are hidden from our eyes.

4. I quote from an article by A. A. Lyapunov and O. Kulagina (*Kibernetika*, no. 16, 1966):

> The Markovian schematics of evolution have the following characteristic feature: Increasing the number of certain self-reproducing forms increases the probability of finding in the next generation more individuals having this form. Regardless of the initial state of the population, if selection acts only on the level of individuals and equally on both sexes, then any deviation from the initial state increases the probability of further deviations of the same kind, i.e., there is positive feedback in deviations from the norm in subsequent generations. Hence the conclusion that when the reproduction schematic is that parent couples with similar genotypes produce offspring easier than couples with distant genotypes, the expectation is that after a sufficient number of generations, a population "polarization" occurs, i.e., this schematic contains prospects of divergence, with fluctuations being subject to positive feedback. In other words, the population's initial genetic distribution may prove unstable. Based on this, one can formulate a hypothesis, according to which a natural trait leading to biological isolation shows a tendency to become stable. Its stabilization is the greater, the lower the number of different states in the given trait. For example, left- and right-handed amino acids do not polymerize with each other. If at a certain point in time, two living forms existed separately based on the two amino-acid types, they would represent two biogeocenoses that do not interact in metabolic processes. But they would still utilize the same pool of the elements, and fierce competition would ensue between the two types of nature. . . . It should be expected that after some time, one of the

forms would achieve victory. Therefore the fact that in the living nature, only one form, the left-handed amino acids, exists cannot serve as an argument in support of one or the other mechanism of the emergence of life. This circumstance is just one of the generalizations of the principle, formulated by Vernadsky,[6] about the impossibility of the retrograde (backward) extrapolation of the evolutionary process.[7]

The article deals with the numerical modeling of the evolutionary process, and it summarizes results of such experiments. A population of 100 to 150 specimens was evolving for 45 to 90 generations under the effect of genetic drifts caused by random fluctuations. The environment was therefore constant, so natural selection, understood as an "adaptation sieve," did not operate. There were three main, distinct outcomes of the experiments: two showed stabilization (either through divergence—i.e., the emergence of several, most often two, species no longer capable of crossbreeding—or through genotypic coalescence into one species), and the third represented what the authors called a "permanent instability," meaning that particular configurations of this state were unstable, but the set as a whole was stable. The first two cases correspond to blind alleys in evolution: the emergence of forms that are no longer genetically reversible, as a result of which a species is "left at the mercy" of the environment and, lacking the genotypic reserves of adaptive variability, it will exist only as long as the environment remains the same. The third represents the preservation of evolutionary plasticity; in other words, a species has at its disposal a variability reserve that is regulatorily necessary, as Ashby wrote in his *Introduction to Cybernetics*.

5. These results demonstrate the importance of random factors in the evolutionary process. It is a characteristic of Markov chains with a finite number of states that if we can define a subset of states such that a chain will have a high probability of transitioning to it but a low probability of exiting it, then after a sufficient number of steps the chain will almost certainly become a member of this subset. Such subsets are called absorbing. It is probable (see A. A. Lyapunov) that the large Mesozoic saurians found themselves in such an absorbing subset, which is why they went extinct. A species can also survive in the absorbing subset, as long as the environmental fluctuations are random and do not exceed certain limiting values. The process of

cultural development should then be recognized as evolution in the Markovian (stochastic) sense and understood as the random walk of a community that can either preserve for a long time (but not forever) the internal variability that enables a continuous increase in complexity (an example familiar to us is the rise of industrial civilization) or encounter absorbing subsets of stationary states, which corresponds to the freezing of some communities at a lower stage of technological development.

6. The boundary between a random fluctuation and an evolutionary regularity (a "gradient of progress") appears to be quite fluid. The reason is the positive feedback between the consecutive departures from the initial state in subsequent generations. By the way, the same feedback exists in inanimate nature. For example, an increase in glacier mass due to a purely random fluctuation (two or three consecutive winters that happen to be colder than average) causes an additional increase that is now nonrandom. The reason is that the accumulation of ice that could not melt during the summers and return the glacier to the state before the fluctuation induces positive feedback: more ice leads to still more ice, and this gradient, manifested by the glacier's descent into valleys, persists until the next statistical fluctuation in the opposite direction (a few unusually warm summers), when the glacier retreats. Similarly, a purely random fluctuation in a population that results in a higher number of individuals with a specific trait, leads to an increase in the number of such individuals in the next generations. The trait does not have to confer any adaptive advantage; it can be neutral in this respect (i.e., biologically harmless).

7. This mechanism could be termed an accident growing into a regularity or a random independent variable turning into a (stochastically) dependent variable, and according to Lyapunov, this explains the diversity of life, which, intuitively, biologists have long considered excessive with respect to the classical dual engine of evolution—variability, adaptively filtered by selection—in the sense that the observed diversity is greater than would be expected if the differentiating, species-forming factors were limited just to the Darwinian dyad. I am talking about the so-called genetic drift, that is, the differentiation caused by intragenetic processes whose results are not actively regulated (through some

limitations, for example) by the environment, because they are neutral vis-à-vis the environment. In other words, the evolving complex system may have certain margins of freedom in which random configurations can be realized and turned into an (orthoevolutionary) regularity,[8] where an introduction of a mutation into the picture does not significantly change it as long as the frequency of such "genotypic innovations" is sufficiently low.

8. Differentiation in primitive cultures, which is "redundant" with respect to the environmental (climatic-geographic) and the instrumental (societally realized technological activities) factors, could have risen in a similar manner and this redundancy or superfluousness can be explained with a stochastic model. A culture's material basis does affect its structure but is not a determining factor; it only forms a space in which variation can be manifested according to the Markovian game of its elements. Biological differentiation due to genetic drift starts with concrete traits already present in a population's genotypic distribution; likewise, societal differentiation starts with basically stabilized relations that can depart from the actual state, branching, increasing in complexity, and "random walking" in the configuration space of possible states. Of course this space is entirely different from that in bioevolution, but the point here is not to reduce the societal type of transformation to the biological but rather to explain the dynamic mechanism that is common to both. It is possible that the crystallization nuclei that "ornamentally" and, in culture, symbolically and signifyingly created intracommunal relations also included, in addition to the relations of cooperation, the relations between the sexes, since procreation and meeting basic needs must have been processes that the evolving group carried over from the biological, precultural realm into the beginning of socioevolution. Differentiation originating from within (i.e., whose driving force does not come from the "game against nature") is in both bioevolution and socioevolution bounded by the initial distribution of the elements (genetic and precultural) and the conditions imposed by the environment, which must be met as a *conditio sine qua non* for survival. Survival is of course not guaranteed: if the conditions are not met during the system's evolution, its direction of the biological or cultural development can lead to self-destruction.

9. The Markovian scheme presupposes a finite number of possible states, but we do not know whether the tree of bioevolution or the tree of cultures actually "tried" all the forms possible. Numerical simulation cannot provide an answer, because the complexity of phenomena that it can access (given the limitations of our knowledge of biology, of machine memory, and of our programming skills) pales in comparison with that in the real world.

10. So the answer to the question of why in some societies "Spartan" ethics operate while in others "Dionysian" or "Apollonian" dominate, why some groups subordinate individuals to the group structure while others, more liberal, value the individual more than the group as a whole, why cultural models of personality are sometimes marked by kindness and other times by cruelty, and why behavioral patterns sometimes privilege and amplify the expression of emotions and sometimes suppress it as reprehensible is that those particular results—after a very long series of steps in a Markovian process—were chosen from many throws of the dice in the "game of society" by random factors that elevated them to the status of the rules of the game.

11. This leads to the conclusion that the plasticity of "human nature" is principially directionless, and in order to form a social group the members must meet conditions that are necessary for the group's stability but insufficient to explain the existence of multiple ethics. An external "ethical selection" factor cannot be ruled out. One can assume that a society's passage through a "catastrophe bottleneck," such as a famine, an epidemic, or other autonomous environmental perturbation, will amplify the principle that man is a wolf to man, granting ruthlessness, guile, and competition a status of rules necessary for survival. But assuming a causative event is responsible for the selection of a cultural feature leads us to a methodologically interesting dilemma that is specific to Markovian models.

12. Establishing a "pure" Markovian path for bioevolution is nothing else than acting in accord with Ockham's razor. We cannot rule out the participation of an environmental factor in the creation of a new cultural feature, when this factor disappears afterward and no research can uncover it. All we can do is show, by simulating the evolutionary processes, that the species-forming divergence may occur in the absence of

such a factor. Which is not the same as saying that this *indeed happened in each particular case*; because for no feature we can say with certainty if it was stabilized with the participation of the environment or arose as a result of a "pure" fluctuation that was subsequently amplified by Markovian feedback. Obviously by increasing the complexity of our model, that is, by introducing environmental factors with various fluctuation frequencies, we can obtain a considerable number of bioevolutionary trajectories. But if we find that the same feature stabilizes in the population, say, 35 percent because of "pure" Markovian selection and 20 percent, 35 percent, and 10 percent because of three external factors that were manifestations of environmental fluctuations that left no trace, which explanation should the researcher choose and on what basis? Gibbs noted that retrospection in sequential probabilistic processes is treacherous.[9] When we are dealing with an ergodic process, which "covers the tracks" of the particular path it took and a given state can be reached through very different sequences of past states, no amount of modeling can determine the real (diachronic) trajectory of the phenomenon. Research can only establish a set of possible trajectories and accept the indeterminacy of what actually happened. This uncertainty differs from that in quantum mechanics, because the process here was definitely not separated into a "fuzzy" linear combination of many paths; it took just one path, only we cannot know which.

13. The hypothesis of ethics arising as a result of an accident becoming a stereotype (or a deviation from the initial state turning into a regularity repeated through generations) is prima facie methodologically "better" than that of ethics arising as a result of a "disappearing" cause (e.g., the passage through a "catastrophe bottleneck"), because it offers a mechanism that is more economical, requiring fewer postulated factors (Ockham's "entities"). Yet the disappearing-cause hypothesis is easier for the humanist to accept, because it argues that something external to human beings is responsible for the creation of a particular ethical system. But of course if we accept it, we get to a dilemma: do "more humane" ethics arise in the absence of perturbations (catastrophes), that is, indeed *anima humana naturaliter bona est*, or do those ethics require the presence of a "positive-ethics selection" factor? These hypotheses cannot be entertained in such simple forms,

however: a linear distribution of the "ethics-selection" factor (on a scale between lack and excess, between the "cruelty" and "kindness" of the environment, between misery and bliss) does not capture the multidimensional "spectrum" of the ethics actually observed in primitive cultures. Because ethics cannot be unequivocally attributed to a causative environmental factor, and we therefore cannot reasonably claim that "humane" ethics arise in "better" living conditions (and vice versa), we are back to the model in which continuous ethical systems arise due to interference of discrete perturbations coming from the environment, which again reveals the Markovian nature of the phenomenon and makes ethicogenesis a probabilistic walk of communal customs through a series of consecutive states until a state that happens to be stationary, that is, "absorbing," is reached.

14. A Markovian process is principially devoid of memory and represents a manner of "learning" that is highly uneconomical. Indeed, historical memory in a primitive culture operates without precision or certainty. Ethicogenesis in it was such a slow process that society could not perceive the manifestations of the very effects that formed its behavior. As we saw, identification of a stochastic mechanism that stabilizes cultural patterns is questionable, but if the environment participates in the generation of the Markov chain, that participation cannot be reduced to a simple scheme where good conditions create "good" (humane) ethics and bad conditions "bad" ethics. The location of the stochastic generator is unimportant here, and so is the "ethical minimum" (which we barely touched in our discussion), which can be reduced to the principle of a group cooperation that enables biological and later societal survival of the primitive community and in which first instrumental activities appear. The stochastic generator for us is simply a mechanism that randomly selects, from all the possible elements of human behavioral patterns, those that form an integral unity to which the members of the culture assign special significance. This process has both a physical and a semantic-cultural aspect. Studying the physical aspect, that is, revealing the purely structural relationships in the formal models of cultures, is analogous to linguistic research, in which we attempt to *replace the understanding of a language* with its grammar and syntax, which can be algorithmically formalized.

15. Views of history range from the idea that it is a fundamentally directionless sequence of states, devoid of regularities (trends, gradients), to the idea that it is a developmental flow with distinct teleological regularities. These opposing ideas can be reconciled if we recognize that the flow of history is not a homogeneous process. There are at least three kinds of processes, variably interconnected: Markovian, cumulative, and random. The Markovian, processes with only a "single-step" memory, are the transitions from a biological species to a culture-forming one. The process of human socialization differs from that observed in animals in that information must be transmitted extragenetically: because at birth, ants, but not humans, already have in them a preprogrammed plan for a societal structure. (In this sense, the species *Homo sapiens* is regulated by two channels: genetic messaging and cultural messaging.) The evolution of societal systems, which depends on (non-Markovian) technoevolution, is also Markovian. Although the process changes its memory from a "single-step" to a deeper kind after the development of an alphabet and historical chronicles, the regulatory effects of that memory on the probabilities of transitions to the next states are quite small. Until the emergence of the theory of socialism, that memory has not been effectively used for regulatory purposes, so from the physical point of view the process has remained Markovian: a memory not utilized does not exist (is not functional). Therefore—and this is methodologically important—technoevolution exhibits behavior more directed than a Markovian process because a continuous accumulation of achievements (learning) is taking place in it. This evolution has a "regulatory memory," although its effects on the Markovian sequence of organic transformations are random in the eye of an observer outside this Markov chain. This is a special case of a general phenomenon: if we have two loosely (e.g., stochastically) coupled systems that do not share their regularities, and if what in the first system is regular affects the second system, its causal effects will be considered random in the second system, because the second system has its own set of regularities which cannot predict the interventions. For example, two cars run side by side, and the behavior of the first driver, which results in their collision, is in the eyes of the second driver random, although it was caused by a characteristic regularity present

in the first car—the first driver has slow reflexes. Therefore, whether a series is considered random or regular can be, to some extent, arbitrary, depending on whether or not the observer is a part of the given sequence.

16. Apart from such mass aspects, history has also a singular one, known as the notorious Great Man theory. When transferred into the sphere of cybernetics, it reveals its indetermination, because it is the system's governing characteristic that decides whether someone "governs" it; moreover, "to steer or govern" and "to regulate" are not synonyms. A driver steers a bus (and holds the passengers' lives in his hands), but the queen bee governs nothing in the beehive, although her presence is essential for the hive's survival—hence her influence is regulatory but not steering. Further, the extent to which a governor's personality can affect the dynamic trajectory of a system depends on the system's structure: some systems will "amplify" the personality; others will suppress or entirely eliminate the individual variability of the ruler's character traits. A governor's actions also can be representative of the system and maintain the values of systemic parameters within the beneficial range without the need of any special talent. But a configuration of conditions can arise in which acts of steering will be random with respect to the system, that is, unpredictable from the viewpoint of its mass-statistical regularities and devoid of any regularity that can be deduced from the overall dynamics.

17. The proposed three ways of describing the dynamics of a society could be further multiplied. When we face a sufficiently complex system, with various "subsets" that are nonuniformly coupled with one another, we can select the description that maximizes our knowledge about the system (and its future states). Different descriptions may in fact be complementary to each other. Some descriptions, especially those that smuggle in "noninstrumental" valuations, are epistemologically inappropriate, as are those that lead to comparisons between, and affording equal treatment to, phenomena that operate on different levels or are of different kinds or use nebulous analogies (e.g., a biological system as an analogy for the too-well-known "parallels" between a biological and a societal system). The three kinds of variability, which can be detected in the flow of history, are not easily

integrated: the leading role is sometimes played by statistical processes (the same kind that statistical mechanics studies in thermodynamics, for example), sometimes by cumulative and teleological processes, and sometimes by singular processes. Therefore a historian's language is usually a mix of at least three different languages—due to the interlacing of the three aspects, which actually operate on different levels.

18. In the modeling approach that takes a system's structure as given, this structure corresponds to something like a country's network of roads understood in technical and technological terms, while ethics, or more generally—cultural norms, is the traffic code, that is, the full set of rules of "proper" conduct for drivers, and the roads and the code variously interconnect. The traffic code must accommodate the real state of the road network; if the code were "unrealistic" or downright unrealizable, it would cause a divergence between the theory that demands and the practice that is observed. We all know how quickly, because of the increase in car ownership, the rules of the road become obsolete. Change that is dynamically similar occurs when ethics fails to keep up with technological innovation. The traffic controller knows that even though the drivers are people, the rules that govern the behavior of large number of cars on the roads after a threshold of "density" has been reached, reflect less and less the element of individual psychology and more and more something like molecular kinematics. The controller is familiar with the phenomena of pulse jams and propagating waves. In such circumstances, appeals to the drivers have little effect even if all those individuals understand the problem and are willing to follow the rules. Once a vehicle ceases to be a "molecule of movement" whose trajectory can be interpreted in terms of psychology, all appeals to conscience will be useless; it becomes necessary either to change the road system (by increasing its capacity or building anti-collisional, ramped interchanges instead of intersections) or to institute new traffic rules, which will have a peculiar consequence if the road capacity is limiting: the new rules cannot avoid discriminating against a fraction of road users.

19. Attempts to justify the principles of ethical behavior have been based on various authorities: transcendental, logical, utilitarian, even psychobiological. The neopositivists eventually concluded that ethics

is nonempirical, because, as Carnap noted in the 1930s, the sentence "Murder is bad" offers no consequences falsifiable by experiment: after a murder, the corpse is evident, but the "evil" is nowhere to be seen.[10] It is surprising that Reichenbach, who worked in the area of probabilistic laws, albeit only in physics, also supported that thesis. Had the neopositivist philosophers turned to technology instead of physics, they would have noticed that there are no true or false machines, only good or bad ones, or rather better or worse ones. "Good" in this sense is a machine or any material system that meets certain criteria of purely instrumental valuation. Technology also has binding directives, and they are a consequence of accepting such criteria. It is therefore possible in principle to compare the value statement of the railroad engineer that "Train crashes are bad" with the ethical valuation "Murder is bad" because they are isomorphic. But both have the same problem: they introduce a criterion that is neither right nor wrong; "bad" denotes only a state that should be avoided, since trains on rails, like people in society, should move without collisions. Ethical evaluations are supposed to differ from instrumental evaluations by being unjustifiable, but this difference does not hold in reality. No doubt the engineer whose train derails at a switch is confirming, among other things, certain laws of physics, owing to which kinetic energy converts to heat, deforms the wagons and locomotives, and so on, yet he is not exclaiming, "Physics is true!," but rather, "The switch is bad!," that is, faulty or poorly constructed. So there are empirical quality tests of material systems that are not identical to tests used in physics. We all agree that the laws of physics are independent of the nonphysical opinions of the people who work with them, but if a locomotive engineer also happens to be a guerilla fighter, he may hold the view that "A train crash is a good thing." That may be true, except that he is then trading the instrumental directives of his technology, with their implied value judgments, for directives that are not technological. Similarly, the "social engineer," who treats a society as a complex machine (in the cybernetic sense), can valuate it according to the corresponding instrumental criteria as "better" or "worse" than another society, either as a whole or just in selected parameters. In his eyes, ethics reduces to a "cooperative minimum," without which no society could function, for a society in which everyone deceives, kills, or robs everyone else cannot exist. Ethics in societies operates as a probabilistic rule (or rather a system of such rules), manifested—in a purely instrumental approximation—as an average of a very large number of individual processes, like the temperature of a gas. One cannot go from this integral "ethical mechanics" to the ethics of an individual, just as one cannot use statistical mechanics to define the temperature of a single atom.

20. The applicability limits of this approach lie where sets of atoms cease to be homeomorphic with sets of people because the latter are a special case of systems whose rules depend on their histories. Here having a history means that the future trajectory is affected (probabilistically) by the past trajectory. Indeed, if the rules of atoms' behavior depended on their history, there would be no fundamental difference between a set of atoms and a group of people. But since the set of atoms is a system whose elements have been indelibly preprogrammed "once and for all," it is a limiting case in the distribution of rule variation, which spans from total, agenetic determinism through Markovian systems to the teleological and diachronic. And vice versa, human society can be considered as a system of particles whose rules are a function of time; we are atoms with memory and the ability to learn—though we have not yet made the best use of that ability.

21. Another special aspect of ethics as rules governing human sets is the selection of the "proper code." Formulated in terms of our modeling, the question is: Is it possible to equate a certain "appropriate ethics" with a class of dynamically "optimal solutions" in a purely instrumental sense, or must we resort to subjective experience and such terms as "conscience," "honor," "compassion," "sympathy," but also "aggressiveness," "death instinct," and "hunger for power," and so on, to describe ethics both synchronically and diachronically (i.e., in its operation and its development)? The answer may come from the simulation of societal phenomena: it is not necessary to assume a priori that the material from which the societal machine was constructed played a decisive role, meaning that a society is as good or bad as its members, that the system just amplifies "human nature." It is even possible that the material is of no importance at all, as in the model

of the brain, as long as the model meets certain simple requirements (e.g., pseudoneurons having two distinct states). Someday we may be able to simulate "sociogenesis," first starting with "molecules that are immanently good," and then with molecules that are "immanently bad." My guess is that it will make no difference, because sociogenesis is a process that is ergodic with respect to its initial state. In other words, a societal system is independent of what is "good" or "bad" in human beings. Imagine a network of roads on which we release a swarm of drivers instructed to be as aggressive as possible toward the others (not observing the right of way, not yielding, etc.), and then the opposite, to be as courteous as possible to all of their fellow road users. Undoubtedly, in the initial stages of the first experiment there will be many more accidents than in the second, but once a certain degree of saturation ("traffic density") is reached, road safety will stop being a matter of personal behavior: physical rules will overtake the "ethical settings." So although different paths are taken to reach an equifinal state, at different costs in terms of accidents, the end result, especially with statistical averaging, will be pretty much the same.

22. The relation between the "ethical" and "physical" dynamic aspects of a societal system can be modeled also in another way: the existing models take individual psychology into account to different degrees. Some minimum psychology must be taken into account, because a society cannot exist if its existence is not in the interests of its members. But requirements of different structures are different and most probably they all feature redundancy with respect to the personal integrity. In order to model these phenomena, at least two alternative (discrete) states of societal elements are needed, Markovian and non-Markovian, that is, without memory (hence with regulatory autonomy) and with it. But a nonbinary approach considering those two states as scale limits within which states can vary continuously would be closer to reality. When tyranny is absolute, all elements are "devoid of memory" because the only thing that counts is regulations, directives, orders of the day; when a system is "ideal," individual memory operates in full freedom. The former is linear, the latter nonlinear. We could then distinguish among various types of tyranny: one type may stabilize the structure by physical force, another by informational

means—the former stabilizes the whole by brute force while the latter by a high effectivity coefficient of internalization of the informational directives; the former is like a military occupation and the latter is something like the Jesuit order. A biological organism is—despite of what we sometimes hear—a curious mix of the two. But if organisms manage fine in nature, tyrannies tend not to last without an aid of certain procedures that, luckily, are so far instrumentally unrealizable, since people cannot be turned into elements that are a hundred-percent Markovian (devoid of memory), as the cells of an organism can.

23. The above model is simplified, as it does not take into account the residual memory that people retain in tyranny. Tyranny is, so to speak, organically incompatible with the biologically determined human nature, in that tyranny does not allow room for the regulatory function of biography, character, abilities, skills, and so on. Even a model that takes into account these "local" properties would still be incomplete, remaining "acultural": in such simulations, cultures would be "energetically similar" but "informationally different" states from the set of all possible states that can be realized in the given systemic structure (for example, many traffic codes can exist for the same network of roads and the same type of vehicles). This would be the next step in our modeling, yet still just the second approximation, because it does not include a feature of a culture that, like ours, uses the memory of all the cultures that preceded it and the knowledge of cultures that have ever been observed, even in lands that we consider exotic. With such a method of successive approximations, we can build dynamic models that approach reality.

24. Yet do not such models omit the specific values to which ethical phenomena can be distilled? If we are studying—say—the struggle for power among the ruling elite, can we neglect personal attitudes, motivations, and intentions? Certain people may derive satisfaction from occupying a privileged position, but the society modeler can safely dismiss that satisfaction, along with the so-called immanent evil of human nature altogether. The modeler will be like an oncologist who calls a tumor malignant but it does not mean that he ascribes to it any malicious intention or wonders if the tumor gets satisfaction from attacking healthy tissues and destroying their local autonomy. The

doctor's task is only to combat such deviations from the organism's homeostasis.

25. The simulation experiments discussed above have not yet been conducted, so we can only assume that as the "rigidity" of organizational relations among people increases, the effect of preprogramming in an individual's behavior weakens. We do not come into this world with ethics, only with the ability to respond emotionally. The newborn responds to a smile with a smile, and that forms the nucleus of the so-called higher emotions, which are plastic in childhood, a time spent, in practically all cultures, within the family. The family is where the first "ethics lessons" are given, in parallel with the acquisition of language, and those lessons are later extrapolated to larger circles of people, the process becoming more and more determined by the culture (family relations being relatively least affected by influences of the culture as a whole).

Yet just as a child's memory is not adequate for the demands placed on the individual by the system structure, the form of communal memory that we call culture may be inadequate vis-à-vis the overall systemic changes in the human ensemble which can be initiated by, among other things, technology. As the rate of evolution increases, the ensemble behaves as if it has lost its memory: it becomes Markovian, and its future states depend only on the present one. The "childhood" preprogramming has an anti-Markovian effect, resisting the memory loss, but the extent of that resistance depends on the culture that formed the family (we are dealing here with a hierarchical series of feedback loops: parents teach the child what they learned about "ethics" from their parents). In this individual aspect (but only in it) personal characteristics should not be omitted. In this way the tendency to manifest ethical behavior arises, which I would call "weak local interactions." By "local" I speak not of physical distance but of situations in which the acting individual represents himself rather than a larger group (a class, an army, an institution, or a government). But no kind of "resistance" can free the individual from the "strong interactions," which are determined by his membership in the larger groups. That is why the noble dream that to prevent a war it would be enough for all mobilized men to refuse to obey the government is utopian. It has never happened; the relations that govern the system cannot be "corrected" by appeals to the heart, as Marxists have known for a long time.

26. By the "technology of ethics," I have in mind specific simulations of ethical phenomena by technological means. But just as we cannot simulate emotions separated from their neuronal—or pseudoneuronal—substrate to obtain "sadness" or "nostalgia" in a test tube, so we cannot simulate what is ethical separated from society. Then a simulation can aim either to model particular processes (say, the "Markovian" formation of ethics in a primitive society), or to produce results that would not be epistemological but instrumental. In either case, the question is whether the simulation allows for a homeostatically rational selection from among "a variety of ethics." An extreme version of this approach considers ethics to be like a traffic code, a set of rules that are impossible to deduce from the description of the material state by pure logic, yet originating from certain instrumentalisms in the form of a multidimensionally optimal solution. Like the traffic code, ethics is supposed to minimize the number of collisions, and, what's more, it should do this "naturally," that is, in a way that benefits all while at the same time not inconveniencing individuals to the point where they begin breaking the rules.

27. But can a rationally thinking humanist "technologize" his love for the common good so much? Calculations show that peacefully united humanity is not just "a good thing in itself" (the ineffectuality of this statement is hopefully obvious) but it also represents the most efficient and dynamically stable system with the highest resistance to disturbances—possibly even at a cosmic scale. Hence an ethics supported by an instrumental, economic, and informational calculus is precisely the one that the humanist would choose. But cannot someone argue that an ethics with divisions, segregation, violence, and exploitation could work equally well in a purely operational and instrumental sense? It is reality, not I, that says that on a purely instrumental basis we cannot put an equals sign between the two ethics. The instrumental arguments in support of humanistic activities are ineffectual in stationary cultures that are governed by either general kindness or cruelty because it may happen that the stability of the overall equilibrium is the same in both. And purely humanistic arguments are disallowed in

my operational approach, not because of the fashionable obsession with cybernetics but because they are culturally relative. If what is good in behavior coincided with what guarantees—metaculturally—efficiency and stability, we would have an ethical compass that would serve even in the most drastically changed future.

In a technologically oriented civilization, antiegalitarianism cannot have the same weight as egalitarianism. First, and trivially, the power that human slaves can supply is nothing compared with the power available in nature. Next, states based on division ("we" vs. "they," "higher" vs. "lower") are always unstable; even if we don't consider the inevitable social antagonisms and the reliance on the use of force, structures stabilized by those means survive no longer than the interval between one industrial revolution and the next. For example, the new era of information, initiated when transmission satellites were put into stationary orbits, makes the total information blackout at any place on the planet practically impossible (in a technical sense). Thus as information-transmission technology progresses, it becomes more and more difficult to keep people uninformed. Economically, a civilization that masters the harvesting of energy from its maternal star or nuclear fusion but at the same time maintains the privilege of private ownership of that technology is—in purely instrumental terms—acting irrationally: it will need to combat an increasing number of difficulties that are unnecessary because they will disappear as soon as the principle of private ownership is abandoned.

True, calculations will not yield this result if they are limited only to short time intervals. Technoevolution cannot substitute for summary justice that punishes the bad and rewards the good—even though it appears to do precisely that in the long run. A homogeneous tyranny might freeze progress by employing brutal but at the same time technically refined means and acting globally (though its existence could be endangered by the exhaustion of the resources behind the technology employed to "freeze" the society). But the reality is that technoevolution is accelerating, which increases the weight of purely instrumental calculations, turning civilization more and more clearly into an energy-information machine whose global equilibrium is increasingly dependent on local equilibria. Therefore, from the global perspective (favored by technological integration), ethical behavior is at the same time rational, in accord with the developmental gradients, because any other behavior will sooner or later destroy the underlying societal order. Such destruction, we might add, could claim the entire species, a global finale to local shortsightedness or plain stupidity.

28. In part I of this essay I discussed the harm to social values caused by some technologies. The question arises: Can technology also act as an ally or amplifier of ethics, hence as an optimizing regulator of interpersonal relations? It is possible, within limits, because technological means can have moderating influence on the interactions between people. A purely physical example would include appropriately oriented production, construction, and other technologies that can alleviate the modern problem of overcrowding, which is evident in shopping, traveling, and housing. But technological solutions do not work without conflicts. In prosperous countries, the problem of pedestrian overcrowding has only transitioned to the problem of automotive overcrowding: more roads need to be built, but this is not possible even in the most prosperous countries. But what appears possible is isolating each person from the others by technology-based "distancing" or encapsulation that allows one to maintain individual dignity (not always easy to do in a crowd) and prevents friction with others among the inconveniences of everyday life. You could argue that technology, by keeping a person's morality from being tested too often and simply making it not worth a person's while to be a brute or swine, is not an ethics amplifier but more a shock absorber in human contacts, prophylactically removing the possibility of a conflict escalation. (Obviously such devices would lose their moderating effect if they were for a privileged few instead of everyone.) If used properly, this technology is a perfect moderator, ethically neutral because it only prevents evil acts. An owner of a home and a car may not be a better person, in ethical sense, than a homeless pedestrian—the former just does not have that many situational opportunities to act in a morally questionable way (though he can of course find them, if he wants to). But surely, any little prophylactic function of technology would be not a bad thing.

29. I cannot imagine how technology could help people internalize morality. Yet in the form of social engineering, it can stabilize a

society's equilibrium in such a way that the behavior of its members becomes irreproachable—externally. And the simple fact of belonging to that society might contribute to the internalization of morality.

The critical points in a societal structure where some people may be harmed by others are those at which one individual becomes the victim of another's freedom. It is not difficult to find such points. But, again, technology can reduce that possibility. (What may be a common dream of modern times is a totally automatic administration, which would turn alienating bureaucracy into an efficient machine.) A solution could be not a personal capsule mentioned above but a kind of filtering distribution system operating throughout the society, faultlessly directing the right people to the right positions, objectively determining the criteria for professions as well as the conditions of work and salary, acting as a universal regulator free of favoritism or malice. In this way we can obtain a structure in which an individual would be protected not only by his personal "technological environment" (home, car, office) but also by the societal system of choices guiding his life's path. With technology acting as a barrier that selectively disallows certain (ethically reprehensible) interactions, an ideal society (from the engineering viewpoint) could be constructed in which we do not have to "do good" to our neighbor because he does not need it—except in rare circumstances (a natural catastrophe, an industrial accident)—and "evil" is not perpetrated on anyone either because it simply is not worth the trouble when it does not provide any benefit (aside from the pleasure that people sometimes feel when they harm someone). I confess that I am not an apologist for this model, although it has merits that contemporary societies generally lack. The reason is that it is built on the hidden (not perfectly) premise of having no faith in people, which is unfortunate, although perhaps rational.

30. The main problem of this model is its stationary character. Only in a culture that is stationary is it unimportant whether behavior results from internal or external pressures, that is, whether it comes from the heart or from instrumental necessity. (The distinction may be important in general but we only deal with practice here, and in practice, morality that results from a drill or a custom is usually indistinguishable from morality that results from the love of virtue.) A

nonstationary culture must ceaselessly adjust its fragile equilibrium, given the continual emergence, growth, and transformation of its institutions (in production, education, distribution, etc.), and it is hard to imagine that a technology of the "ethical neutralization" of interpersonal relations could keep pace with all those changes. Ethical values, once internalized, cannot be changed. But the premise of no faith in people requires foresight to keep new inventions or technologies from being abused. However, the unequal development of science and technology, as well as its long-term unpredictability, makes the full success of any such safeguarding impossible.

31. I have focused here on the search for technological ways to protect humankind from itself, that is against actions that may be fatal for individuals, communities, societies, or the entire species. I note that the less numerous the group in question, the more difficult it is to establish instrumentally the rationality of ethical behavior—at all levels (from a family to a country). The most difficult, perhaps impossible, is to prove the irrationality of misdeeds of individuals—we know how many rogues go unpunished. The increasing success of forensic technology in identifying perpetrators is not a good argument in support of society growing more virtuous; it can also be an argument for the need for criminals to improve their skills.

32. Technology as an aid for ethics can accomplish much in the field of social engineering, if only by introducing "dampers of evil" into existing structures or by making gradual innovations. The ideal would be a structure with three features: societal impermeability of the negative actions of individuals or groups; transmissibility, with amplification, of the positive actions; and, between these two, the maximum number of degrees of personal freedom. This "smart" structure need not be made of "smart" components—by "smart" I mean features that a living organism possess: the ability to repair itself, ultrastable equilibrium, and energy efficiency that does not depend only on the "smartness" concentrated in the nervous system. Bioevolution would certainly not be possible if adaptive success depended on whether or not each animal "figured out" that to survive it must breathe or use this protein and not another to counteract a bacterial toxin. As we know, it is possible to have the brain of a chicken and still prosper,

but as of now no governmental organism can be run by managers with the intelligence of a chicken and flourish politically and economically. The obvious drawback of societal structures is that they act rationally only when they have rational managers. The relevance of this random factor, which today cannot be avoided, could be decreased by an appropriate reorganization. Bioevolution formed only "good," that is, "rational" structures because it had enough time at its disposal, in which even a Markovian process without cumulative memory could find dynamically optimal solutions. In contrast, the period of trial and error in the construction of societal systems, lasting barely a few millennia, has not yielded such success.

33. Moreover, those trials did not include any theoretical planning, which, as we know, can tremendously accelerate progress. Nevertheless, the question arises whether such a perfect "sieve" that permits only the "good" to pass and blocks the "bad"—provided it can be created out of human atoms—is worth realizing and would even be possible to realize socially, regardless of any technical issues. There are three kinds of difficulties on the path to the modeling of societal phenomena, which must precede its realization. The first is formal and technical: the selection of a language (or languages) of description, essential parameters, and quality criteria for the obtained results. A particular problem here is how to make the "optimal structure" invariant with respect to all the unpredictable transformations that the future technological revolutions will cause. What is easiest to simulate—a rigid development of the orthoevolutionary type—is as trivial as it is useless, whereas the more interesting systems, with high levels of complexity and individual freedom at the same time, are nonlinear. Thus what is most predictable, due to its clear regularity, is not worth realizing and what is worth realizing is hard to represent statistically. And there are many other similar problems. For example, large system aggregates may pass through a series of developmental critical points, in which the effect of a random parameter dominates the internal feedback regularities (i.e., the system suddenly becomes oversensitive to local fluctuations), in which case a great many alternative paths will open up as potential radiations. Randomness can certainly be simulated but only in an overall, not particular, way, and when the number of possible

solutions exceeds a certain limit, the problem will become impossible to solve, even though solutions may theoretically exist. On the other hand, the orthoevolutionary tendency of a technologically oriented culture will facilitate the simulation. Future experts will have to consider all these issues.

The second kind of difficulty is the widespread attitude that the whole endeavor is impossible. This is slowly changing, but a climate of support is lacking, and so is the understanding that meaningful research in this area will require large teams which must include anthropologists, sociologists, mathematicians, and others. It is high time we recognized the fundamental priority of this field. It needs appropriate channeling of efforts, inflow of investment, and an orienting of the best minds, simultaneously brilliant and properly trained, in the right direction. A related difficulty is political opposition to such a project. Imagine that a team of researchers presents a solution to the Vietnam problem that contradicts the United States military doctrine. Any dialogue between the team and the government, not to mention a fair consideration of the proposed solution, would be out of the question, since the strategic doctrine of the United States implies that the possibility of "losing face" is unacceptable, whereby a superpower's fear of being embarrassed outweighs the fate of the species. It is equally clear what would happen with a mathematical discovery on the topic of, say, operation of selection filters in the power elite circles.

The third kind of difficulty regards the principial uncertainty of the simulation results. In an "optimal" culture, people are supposed to be "perfectly satisfied." But in modern civilization, the control of societal parameters can be such that to an external observer a situation will seem to be what, in reality, is the complete opposite. Manifestations of mass enthusiasm, anger, or chaos can be carefully arranged and imposed. Consider the "mess" of the road signs in southern England during the Second World War; it was artificially created with the aim of confusing the Germans in case they invaded. There is nothing easier than to create a state in which everyone claims to be completely satisfied. One problem is that if the pretense is maintained long enough, it may, for all its monstrousness, become a sui generis truth. Some people who lived for a long time in the concentration camps forgot their previous life to such an extent that their response to being liberated was shock, passivity, frustration, even despair, because they feared this new, free life more than the awful but familiar conditions to which they had adapted. Metaphorically, to make the slaves free, it is not always enough to break their chains. We therefore cannot a priori rule out that if the theoretically optimal model were realized in life, it might be like the bed of Procrustes, but people's extraordinary adaptive plasticity would prevent us from finding that out: being stretched on the bed, people would still insist—with sincerity—that their life is perfectly fine, and if there was any discomfort, the fault lay in their own bodies or in their nearest neighbor.

34. Why do I keep repeating ad nauseum that the time is ripe to undertake socioevolutionary modeling, in which computers will play a leading role and enable the simulation, in accelerated time, of societal processes? Because without this effort, which may be rewarded with some success in the next century, the technological acceleration will probably make our planet's equilibrium even more fragile than it is now, and then it will be too late to do anything. All the variants of "physicalization" of our topic that I offered above are hopelessly primitive, but there can be no discovery without an agreement on methodology and without plans for work teams of appropriately educated and trained experts, for which I plead. Had it not been for the great mobilization of minds and means in response to war, the task of liberating the atomic energy might have remained unsolved up to this day. The same if not greater accumulation of effort is needed here. No matter how utopian it sounds, I repeat what was said before: "There are still no sociologists that would, following the physicists, demand billions for machines to 'simulate societal processes,' not to speak of ethologists who are today just Pascalian reeds in the world's gales.[11] Yet we ought to have faith that someday the situation will radically change."

Summary

1. I expressed my conviction that the effects of metaethical factors, such as technology, on the formation and functioning of ethical systems, as well as the effects in the opposite direction—of ethics on

what is nonethical—can be studied empirically, in a rigorous manner, and that the results of such investigation, supported by modeling these phenomena in a nonsocietal and nonhuman substrate (such as computers), may provide important directives for instrumental behavior that asymptotically approaches the creation of "the ideal societal structure."

2. In particular, I consider ethics a result of the averaging and embedding in the societal realm of an enormous number of elementary (discrete) acts of personal behavior, which create, on the one hand, moral norms and, on the other hand—by idealizing these norms—axiomatic generalizations expressed as an obligatory-axiological formula. One of the essential parameters of culture may be the degree to which the real behavior in it diverges from that formula; this can be observed only in the form of a statistical distribution.

3. The individual-psychological (experiential) aspect of ethical behavior has been omitted here out of principle, as replacing that form of description with the "physicalized" one greatly simplifies the future simulation of these phenomena. But the omission does not mean that this aspect is insignificant; my approach has been similar to that of medicine, which, in the presentation of a disease, dedicates little space to the suffering (experienced introspectively) that the disease causes, even though the purpose is to remove precisely that suffering.

4. According to the presented hypothesis, what is "ethical" constitutes a part of the regulatory characteristic of the group behavior that has the highest probability to be realized in equivalent situations, a part which—just like the whole group-behavioral programming—is a resultant of at least three things that participate in the stabilization of given behavior: random circumstances (such as climate fluctuation), Markovian processes (which stabilize the results of random deviations from the initial state through positive feedback), and cumulative developments (e.g., technoevolution). These three create an intracultural model of "human nature" and stabilize the related system of norms and ethical values that for the members of the given culture is not simply a collection of probabilistic preferences but carries symbolic significance.

5. Ethics is thus cocreated also by metaethical factors that are essential for a group's survival, are either factually or in the group members' opinion necessary for maintaining a group's continuity, and can be physically realized (a custom of flying or a norm that prescribes flying can never arise due to the lack of physiological mechanisms and instrumental means).

6. I have compared the development of a precultural group with the evolution of a species population, using a Markovian model, and pointed out the similarity in the radiations of variability in biology and cultural anthropology: in both cases, the variability is superfluous with respect to selection factors. Another source of analogy is the existence of stable ("absorbing") states in a Markovian process; the "freezing" of biological and cultural forms at certain stages of development can find justification here. I also introduced the idea that a genotypic "permanent instability," acting as a reservoir of regulatory (and potentially adaptive) variability and thus enabling the continuous evolution of biotypes, may be like the "permanent instability" of a technologically oriented culture, which maintains its relative equilibrium only thanks to the accelerating technoevolution that takes place in it.

7. I presented, in a quite primitive way, a possible "physical" model of society in which one can divide ethical phenomena into "weak local interactions" and "strong nonlocal interactions" which arise in and among large ensembles; the strong principially dominate the weak. This picture does not mean that there is no conflict between the norms "molded" by the strong interactions and the ethical, weak norms in individuals and that the latter cannot gain the upper hand in personal behavioral regulation. As a matter of principle, I was interested not in what the members of societal organizations or the representatives of institutions feel when they execute orders but in their objective behavior, and only in statistical average at that.

8. In an example I showed the effect of a narrowly defined technological interference in human "biological nature" on the functioning of ethics (sexual) in a situation that is characteristic of a highly developed civilization. The interference indirectly destroyed values that had been traditionally considered virtuous. The conclusion was that

ethical, moral, and generally cultural predictions are needed before introducing any technology that can change the natural parameters of the functioning human body.

9. I also presented a technological means whose widespread utilization could help individuals, by acting as a (prophylactic) suppressor or filter, avoid unethical behavior. This technology is "ethically neutral," as its operation is limited to "removing the opportunity" for actions that can harm others.

10. It appears that a stationary culture, that is, one that does not significantly change over several generations, can be modeled as a hierarchical whole in which "strong interactions" mold the "weak local interactions" unidirectionally. Feedback effects of the "weak" on the "strong" are negligible, which means that under the influence of accepted ethical norms the main societal gradients are not subject to correction from negative feedback, and that therefore the system structure (which generates these gradients) is insensitive to local ethical interactions. Even though both strong and weak interactions are random dependent variables, the former rule the latter. At the same time, the stability in all cultures of their elementary units, the families, can make parts of the "weak interaction" programs, such as those inherited from earlier cultures, resistant to the influence of the strong interactions. Because of their multilevel nature, these relations are similar to the bioevolutionary schematic of the emergence of a species, where the genotypic and phenotypic factors are correlated and a (Markovian) circulation of information is taking place at different levels (at the genotypic "microlevel" and at the "macrolevel" of phenotypes, i.e., mature individuals). A stationary culture is then ultrastable, as is a perfectly adapted species. Ethical "inventions," "improvements," or "easements" carried out in groups with few members or by individuals usually are not adopted by the society as a whole. Behavioral traits that arise in individuals and are not assimilated culturally are similar to the bioevolutionary schematic in which an acquired trait is not inherited.

11. In a technologically oriented culture, the exponential acceleration in the variability of living conditions often has consequences in the moral and ethical sphere. When the acceleration in variability changes exceeds a certain limit, the intergenerational transmission of

norms (both instrumental and noninstrumental) can break down, as the norms of the parents become obsolete and cannot accommodate the situations of the children. The result may be a kind of a societal drift of values in the stream of perturbations caused by the technological acceleration. (I omitted the phenomena of so-called mass culture and its moral derivatives due to spatial and thematic constraints: they would require too much room to discuss and besides are the topic of numerous works in the field.)

12. I was deliberately one-sided in presenting the intracultural functioning of individuals subject to "strong interactions," as if the "material" of which the society is "constructed" were not significant. I suggested that "personal" parameters, that is, what we normally call character, resistance to stress, intelligence, drive, extra- or introversion, and emotional sensitivity may not find good correlates in the model of a society as a complex system (in the cybernetic sense). I spoke of the irrelevance of the "material" in the extracerebral simulations of the brain processes. This point of view resulted from the assumption that the maximum economy of means (and Ockhamian "entities") is needed to make the simulations feasible. Taking this view does not mean that I consider personal parameters "irrelevant"; on the contrary, I firmly believe that a society ought to exist for individuals, not vice versa. The "reduction" of human individuals to points in a kind of "configurational space" should thus be understood only as an extremely simplified method of description.

13. Finally, I listed the principal obstacles on the proposed modeling path, difficulties that are partly technical, partly epistemological, and partly methodological, because such is the problem of antinomicity or equivocality ("indetermination" or "uncertainty") of the experiments in which the elements—people in the experimental societal system—are subjected to empirical tests that are supposed to determine whether their life in that system is "good" or "bad."

14. I conclude that it is practically impossible to model large historical processes, such as the evolution of terrestrial civilization. The Markovian nature of the phenomena, forbidding backward extrapolation in any distinctly ergodic process, thwarts the endeavor so thoroughly that other reasons for this impossibility do not need to be given. But

this impossibility, which equally applies to the path of the terrestrial bioevolution, does not preclude the modeling of parts of the process. A technologically oriented civilization, because of its markedly teleological character, should be more suitable for this modeling than cultures that are instrumentally primitive. This at least gives reason for some optimism as we face the future.

BIOLOGY AND VALUES

Introduction: Values and Aims

When we speak about values, we understand them either as facts or parts of relations. In the first case, we say that something *is* a value; in the second, something *has* value. The first way of speaking makes the value absolute. If I say that X is a fact, and X is the distance from the Earth to the Sun, I am claiming a truth that is independent of circumstances. Similarly, if I say that X is a value, my claim does not depend on anything else. If a value is a fact in this sense, its confirmation is nonempirical. There is no procedure that can determine whether the statement "X is a value" is true or false. An assertion that establishes a value in absolute terms therefore does not belong to the language of empiricism because it is undecidable: we can neither falsify nor verify a statement like "Virtue is a value" or "The human is a value." On the other hand, if I say that X has a value, I am saying not a whole sentence, just a part of it: because X is some relation—of subordination, suitability, usefulness—with respect to what is not X. The value that X has is given by the extent of its connection to the occurrence of some Y.

Values are autonomous and absolute in the first sense but not in the second, when they are usually called instrumental. Many statements about autonomous values can be reformulated such that they become falsifiable statements about instrumental values (in considering how society benefits from the virtue of its members, we can reformulate the statement "Virtue is a value" to "Virtue has a value"—for example, the value of stabilizing social relationships). Both kinds of statements imply duty: the first does it categorically, peremptorily, in the form of commands regarding attitudes or behavior consistent with

---

First published in *Studia Filozoficzne* 3–4 (1968).

a given axiological thesis; the second does it conditionally and relatively. Because only someone who believes that social processes *ought to* be stabilized will want to implement virtue in communal life; only someone who *intends* to build a house will want to scrutinize instrumental values of various construction plans.

As can be seen from the above, autonomous values are treated as if they were ahistorical facts or even eternal truths. No axiologist would say, "Virtue has been a value since the year 1456" or "Justice was a value between 1366 and 1890." But if someone did say that, he would actually mean "Some people behaved virtuously at that particular time," and it is immediately clear that he committed hypostasis when he objectified behavior that occurred in a specific, closed time interval in the past.[1] Instrumental values are time-dependent and historical: the instrumental value of flint quartz used to be considerable, while today it is zero. If some things seem to have an instrumental value that is unrelated to any technology (e.g., the atmosphere), it is because they have been "permanently included in the technology" of the human organism and are necessary to support life. Yet it is clearly an instrumental value: if all life on Earth were annihilated, the atmosphere would lose its instrumental value, since there would be nothing and nobody to whom it would serve as a supporting medium.

Because the values of the first kind are established and those of the second kind are uncovered, the former, being dependent on resolutions or agreements, are firmly anchored in ontological viewpoints, whereas the latter can be reduced to a form of pure ontological neutrality, which is typical for empiricism. Ultimately, autonomous values are also relative—but only with respect to a *given ontology*: they are what it says they are.

Every axiological argument is in some way connected to teleology. By teleological behavior we mean that a system inevitably (but also facultatively) proceeds from the current state to certain future states, which thereby become aims. Every description of teleological behavior translates into the language of regular causal description, assuming a complete determinism. The reason is that there is no empirical method that can differentiate between what occurs inevitably because it is a casual consequence of a past cause and what is an aim or final state at the end of a defined path that also had a specific starting point. We can say either that material systems proceed to states of higher entropy because disorder increases or that the aim of every system is a state of maximum entropy. Yet using the term "teleology" in the latter case is not appropriate because of the principle (nota bene empirical) that teleological terms are used only in the former sense, and only an aim whose achievement is not tautologically identical with causal determination can be an empirical term. In other words, an aim is real only when the possibility exists that it will not be achieved. If there are processes that seem to progress toward specific final states even against opposing forces of the environment, and reaching those final states can be confirmed by experiments, then the physically measurable deviations from the path that leads to the goal and the distance from the goal are subject to instrumental evaluation. What increases deviation from the path acquires a negative instrumental value; what helps the process stay on the path acquires a positive instrumental value.

Measurable values thus appear where *real* aims exist. By real aims, I mean only those that are not always achieved. A shooting target is a real aim, but the entropic *finis mundi* is not.[2] Axiology can be empiricized only to the extent it can be ontologically neutralized.

I. Axiology and Physics

From Laplace's deterministic point of view, there is no informational difference between knowing the past and knowing the future: both knowledges can be equally perfect, that is, complete. Laplace's demon, possessing such knowledge in terms of actual states of all the atoms, can describe in the language of physics the behavior of stars, amoebas, and human beings, whether retrospectively or predictively, while *totally omitting all axiological terms*. The shooter may not know that he will hit the flying pigeon, and Romeo may not know that he will meet Juliet, so for the shooter and Romeo the pigeon and Juliet are real goals, and it is the difficulty to achieve them that gives them a value. But Laplace's demon knows that the shooter will hit the target even before he takes aim and also knows what fate awaits the two lovers even before they set eyes on each other. The values inherent in human choices and the goals that human beings strive to achieve are fictitious for the demon. The difference that the demon sees between the goals of the shooter and those of Romeo is just that relatively little knowledge about the atomic distributions is needed to predict the outcome of the shot, but a large amount of knowledge is needed to predict the outcome of the star-crossed love. Yet in reality whether the target is hit and whether the lovers meet is not determined by any specific sequence of events that was *initiated* at the shooting range or on the balcony. The cause of these and all other events in the universe is the original atomic distribution in the primeval nebula; everything that happens afterward is the completely predetermined consequence of that distribution. For the demon, possessing knowledge that is ultimate because no greater can exist, there are no other goals and no other values than *fictitious*, because, as I have already said, a real goal is nothing but a future state, the achieving of which is indeterminate. The pigeon is a real aim for the shooter because the shooter does not possess the optimal knowledge. If full determinism rules, then everyone who believes he is taking aim is like a streetcar passenger who thinks he is making decisions about the car's destination. Of course, the destination does not depend on his "decisions" at all, and so his chosen goal is not real. When we say that Mr. X has chosen a certain goal, we understand that he has selected it out of all possible goals because of its value. Laplace's demon uses no teleological or axiological terms in his description of Mr. X. Mr. X is thus like the streetcar running on the rails to all his future states, totally predetermined, except he is not aware of that. The values he is talking about are nothing but a manifestation of his *lack of knowledge* about states of himself and of the environment. Values are then, from the demon's viewpoint, false hypotheses about the causes of Mr. X's behavior. The physicist will know that Mr. X must behave as he does, that Mr. X does not really have a choice, and that his decisions are as fictitious as those of a streetcar going through a preset switch, although he believes that he has a certain amount of freedom to make decisions according to his system of values. In extreme determinism, goals and values are just putty filling the gaps between the consecutive states that constitute a diachronic trajectory. Values and aims are epiphenomenal illusions, like the daydreams of the streetcar that it is not a streetcar at all and does not run on tracks. Physical knowledge is therefore the only possible kind, and at the same time is ultimate, because by filling the gaps in the description of a system's successive states it erases every trace of value or valuation.

But this "streetcar" determinism is itself an illusion, with no equivalent in reality, even for physics itself. Following modern physics, I posit that the complete determinism is just an ideal limit of real states and that the links between those states have a stochastic character, which often may be disregarded in practice. It can be omitted in the prediction of a lunar eclipse but not if we study electron diffraction. Mr. X's behavior, like that of any stochastic system, such as a die in a game of dice, can be described using the mathematical language of probability (e.g., a Markov chain).

A homeostat is a system that maintains its stability despite perturbations. It is therefore a system whose equilibrium is stable, although it does not have the highest thermodynamic or statistical-mechanical probability. The difference between a homeostat and a nonhomeostat is the same as the difference between the deterministic and indeterministic behavior of an object. Descriptions of atomic particles need to take into account their quantum-indeterministic properties to be of any predictive value. As the number of particles increases, such that they finally constitute a macroscopic object, we are increasingly justified to omit the quantum aspects and utilize the apparatus of classical physics. The extent to which we can omit an object's quantum properties depends on its dimensions, which, in turn, depend on the number of elementary particles participating in the "event" of their variably stable "encounter" that constitutes the object, and on the degree of order the particles exhibit while forming the object. The set of all possible arrangements of the particles contains a subset of systems that we call homeostats. But that subset does not have clear boundaries, because, depending on what parameter coupling arises as a result of a given order, the object will exhibit various degrees of stability with respect to various kinds of effects. The more stable its equilibrium and the higher the number of perturbations that will not push the system out of balance, the higher the extent to which the system is a homeostat. Stability in itself is a necessary condition but not sufficient. An isolated, cold celestial body is in an equilibrium that is stable both mechanically and thermodynamically, but it is not a homeostat; any mechanical or thermodynamic action will push it out of that equilibrium, because it does not resist perturbations but "embraces" them. A planet with oceans and an atmosphere will behave like a homeostat in terms of surface temperature, because an increase in insolation causes an increase in water evaporation, and the resulting clouds increase the planet's albedo.[3] As a result, more of the incoming radiation is reflected into space, and the surface temperature does not rise to the degree it would if this parameter coupling did not exist. In astrophysics, we could, in principle, speak of planets as better or worse homeostats; that it is not done is not because it is physically incorrect but only because such thermal compensation work of a planet is inconsequential for its physical evolution. (Unless we study planets with regard to their ecospheric adaptation to produce life: in that case it makes sense to say that some planets are better or worse in their suitability to becoming cradles of biogenesis.)

In a given environment at given values of pressure, temperature, chemical composition, and so on, there exist states of material ensembles that can persist without significant changes with no energy input, but there also exist states that do require energy for that purpose. The ensembles that manifest permanence or "self-preservation" without input of energy are not considered homeostats. A stone block, a cast-iron sphere, a stool, or a diamond are not homeostats even though they can maintain their structural and material identity for a long time. An oil droplet swimming in alcohol is also not a homeostat. The energy input that a homeostat requires indicates that to maintain an invariant structure, that is, to keep the values of a group of parameters within a specific range, requires work. Yet a steam engine is not a homeostat either, because it requires energy that it cannot obtain on its own. A steam engine that could search for fuel and repair itself when it breaks would be a homeostat. It would have to have various information sensors, since searching for fuel requires good orientation in the environment. If instead of the steam engine we built a photosynthetic machine, most of the sensors would be superfluous, and the machine could be stationary, because the sun shines everywhere on earth, albeit intermittently. Any green plant is such a homeostat.

Artificially constructed homeostat models differ from natural homeostats in that they do not function in the full self-preservation range that a regular terrestrial environment requires. The models are narrow-range homeostats, perhaps just quarter-homeostats. These apparatuses model only selected functions of the systemic parameter stabilization, like Ashby's homeostat, whose equilibrium is not truly self-preserving. A truly self-preserving homeostat will respond to attempts to damage it as a dog does, either by fleeing or biting its attacker. Ashby's homeostat can be destroyed in a number of ways, and it will not defend itself, lacking the appropriate structural-functional features.

If a system is a homeostat, some states of the environment will facilitate its maintenance and others make that maintenance more difficult or even impossible. Being beneficial, the former environmental states will have a positive value for the system, and the harmful states will have a negative value. Here is where the theory of homeostasis diverges from physics. A physicist does not care whether the system being studied is destroyed or not; he observes and describes its physical states—with stable or unstable equilibria—but does not assign them any value. He will not say, for example, that stars that spend their nuclear energy more economically and therefore shine longer are better than stars that exhaust their energy quickly. Instrumental axiology arises in the description when the processes taking place in the homeostat start to be compared with a state that is set as an ideal that ought to be preserved.

When a comet's head enters the Earth's atmosphere and, as a result of the increasing temperature, the ice in it forms a gas pillow that has a braking effect so that the rocks from the comet's nucleus land on the ground unshattered, we do not say that the comet acted "appropriately" and "saved itself." But when an astronaut in free fall, before he enters the atmosphere, uses a foam plastic to create a shield that allows him to land safely, we do say that he acted "appropriately" and "saved himself"—because we know that the comet could not have behaved differently while the astronaut could have. What is the operational difference between the necessity of behavior and the possibility of behavior? The "soft landing" of a comet is highly improbable. To increase this probability the comet should be modified. All its frozen gas would have to be shifted to the front. Except that we do not know the comet's orientation at the moment of entry. Also the rate of transformation of the comet's ice into gas is important for a soft landing. Taking all of this into account, specific pieces of the cometary matter should be rearranged in a nonrandom manner. In other words, the comet would have to be turned into a system with higher order than it had previously. The chance of a soft landing could be increased further if we equipped it with a sensor, a radar that could compute its distance from the atmospheric envelope, and the comet would orient itself based on the data from the sensor. The computation could be done by a computer on the ground and transmitted to a receiver on the comet by radio. Or the computer could be on the comet itself. This is how we could restructure the comet into a self-orienting apparatus that optimizes its flight trajectory.

But it is still a deterministic system, albeit one with a probabilistically selected range of behavior. After the comet's soft landing, nothing would prevent people from smashing it to pieces, say, with a hammer. To protect it against that destruction, we would need to amend its landing program and equip it with additional sensors and effectors. Since not everything around the landing site would present a threat to it, the information from the sensors would need a discrimination filter, which means that the "comet" would need a kind of perceptron. After many additions like this, we would eventually obtain an apparatus able to make decisions, on the basis of preprogrammed instructions and personal experiences, to preserve itself. The system would gradually, step by step, change from deterministic into probabilistic and ultrastable. We have just described a series of transformations that change the comet into an object that is increasingly more like the astronaut. Obviously, unlike him, it could not talk or have children—but this is just a matter of additional reconstructions to bring what used to be a comet to still higher levels of complexity.

The questions of when and how instrumental values arise can be answered like this: the difference between the presence and absence of axiology, and equivalently, between the presence and absence of a real goal, is no other than the difference between a bald head and a mane. When a rock falls in a gravitational field, we do not say that it has chosen to accelerate the speed of its fall over time. But when a virus approaches a cell, our interpretation is not as straightforward. On one hand, the virus's behavior is nothing more than ordinary catalytic reactions between large protein polymers, yet we say that the virus is "attacking" the cell, parasitically harnessing its energy and structural material for its own reproduction. We may accept that the virus definitely does not make decisions in an axiological sense, and that a bacterium doesn't either, but we may hesitate at the level of an amoeba or, if not there, at the level of an annelid, or still further up. Because basically it comes down to this: if we understand the model of a homeostat's functioning as completely as we understand a functional scheme of, say, an electric doorbell, then for us the "making of decisions" will be replaced by feedback and causal relationships, and "functional aims" will be replaced by probabilistic chains, which in limiting cases (a mouse, a monkey, a human) will acquire a status of *models* of the homeostat's environment. "Values" will then be specific relations between physical states that statistically determine a system's behavior. What are these relations? They are not material or energy relations, as the relation between a supporting foundation and a supported wall, but informational relations, which are not objects but facts (just as the distance between a prey and its predator is a fact but not an object).

Information in the physical sense, as a measurable quantity, belongs to subjects studied by thermodynamics and statistical mechanics. But how does physics view information in the logical sense? As a representation. And what is, physically, the operation of representation? Suppose a tribesman A eats spoiled goat meat and dies. Tribesman B eats unspoiled goat meat—and dies too. We can describe what happened to the first person in a completely physical way. But what happened to the second cannot be described in the same language, because that language does not have terms permitting the statement that goat meat is taboo for tribesman B, and when he discovered that he had accidentally eaten goat meat, he died from the shock. An objective, physical description would note shock as the cause of death but would not include the relationship between the eating of the meat and the death.

Tribesman B died because he assigned an extremely negative value to the consumption of goat meat. What in the physical description could correspond with the assignment of a negative value—a taboo? It would seem at first that nothing could, but in fact this assignment corresponds with a series of physical events that resulted in tribesman B's being suitably preprogrammed during his personal history. It is a question of coupling semantics with physics. No available physical test may be able to find any difference between tribesman A and tribesman B. Yet the two are not truly isomorphic, since one died because the consumed meat was poisonous and the other died although the meat was not.

The propensity of dying from the consumption of the taboo meat is a trait of tribesman B that we could predict with a certain probability if we knew all his past states and, in addition, all the past states of his cultural group, going back to the first anthropogenic transformations that led to the rise of language. The idea that improvements in the physical description of the brain of a person who spoke a certain language would ever enable us to uncover the semantics of that language is wrong in principle. A perceptron is a much simpler device than a brain, but no matter the level of detail to which we disassemble or otherwise dissect it, it will never tell us what geometric figures it has learned to recognize. At the start of the learning, the perceptron's elements were connected randomly and only chance determined which elements in what configuration were excited by which stimulus; the excitation pattern was random at first and only after a series of repetitions ("learning") gradually became an invariant behavioral feature of the device. Yet there is nothing either in the device or in the geometric shapes it is learning to discriminate that would make all perceptrons that recognize the same shape assume an identical structure. Similarly, there is nothing in objects or in the names given to them that could tell us why a particular word represents a particular thing. The reason is that linguistic representation is *logical*, not *physical*. As we saw in the perceptron example, however, the process that *actualizes* this logical representation is a totally ordinary physical process. The phenomenon is always the same, although at different levels of complexity—in perceptrons, in the communication of bees, in the linguistic anthropogenesis. At the beginning, we have unconnected distributions of possible "designates" and possible "names" for them; through multiple stochastic processes of "reaching out" and "mutual fitting," nonrepresentational randomness turns into nonrandom representations. Because these are probabilistic phenomena that eventually arrive at states that are *stable* (a working language or a perceptron), this is an ergodic process.

We thus arrive at the following picture: As long as the "mutual fitting" between the ergodic of the facts and the ergodic of the behavior (of a perceptron, a bee, or a human) has a purely physical character, it is not a logical representation. When eventually the representation becomes logical, the physical link that actualized it disappears. Meaning as a physical state is determined logically, but it arises through successive filtering of a set of possible names—as specific configurations "fitted" to the set of possible designates. What "supplies energy" to this mutual fitting of the two ergodics and narrows the initial random distributions of events (behaviors) to *names* is a derivative of the process of adaptation. (The naming is only simulated in the perceptron, since the device "is not interested" in recognizing geometric patterns, which is precisely why such a device must be constructed artificially. But the bees are definitely "interested" in having a signalization system for communicating the localization of a food source, and the human group that has undergone socialization is interested in having a code of societal regulation.)

When randomness becomes regularity, semantics arises as an invariant. It follows that the meaning of a "taboo," as with any linguistic meaning in general, cannot be uncovered by dissecting a brain, because it is a search for something that has not had the form of a physical phenomenon (a linguogenic ergodics) since prehistoric times. What we can observe is only the dynamically stabilized effects of causes that are long gone; it would be as if a physicist claimed that a rock falls from his hand not because he released it in the gravitational field but because the rock fell like this before and is just "remembering" that behavior now.

So it appears that "physicalization of culture" will forever remain a utopia. But if it were ever somehow accomplished, values would become "superfluous" entities (*entia praeter necessitatem*) in the understanding of the Laplace's demon.[4]

The following statement by Rosenblatt, the perceptron's creator, is well known: a perceptron whose elements are connected *totally at random* can start functioning (discriminating shapes) right away only if the number of its elements is infinite.[5] Thus only an "infinite perceptron" can dispense with the process of organization. The higher the degree of its initial organization, the fewer elements it will need, and therefore the simpler it may be. (This is called the self-organization theorem.) This enormously important assertion, however, cannot be directly used in biology because neuronal systems are neither simple nor hierarchic perceptrons, and the (genetic) preprogramming of their subsystems is subject to more than a single rule. But the theorem, if suitably reformulated, can be used for the classification of typical evolutionary decisions regarding the construction of systems through the narrowing of their "informational preregularity" into the "limiting range of patterns" that either an individual—using his brain—or a species—using gene populations—can learn to recognize. Speciation too is a kind of "learning."[6] The self-organization theorem can shed light on the limits of information resolution of a brainlike system. I am hinting here at the possibility of using perceptron technology in the realm of epistemology.

The axiological evolution as a correlate of biological evolution is hard to simulate, since what we are currently able to model is usually trivial. One might say that we get from our models only what we put into them, so epistemologically such experiments smack of tautology. We are unable to reproduce self-organization that would turn an axiologically neutral system into an axiogenic one. A homeostat that searches for electrical outlets when its batteries run low has not yet emerged in a kind of electrical evolution; its behavior, determined by the preprogramming built into its structure (the schematic of its circuit connections), is therefore as deterministic as that of the streetcar, except the homeostat carries the "restricting tracks" within itself. A homeostat that would be a true axiogenerator is one that could proceed from a state of axiological zero, which only recognizes ordinary physical objects, to a state of a specific, "self-centered," self-preserving knowledge (of what "needs to be done" to maintain activity). A homeostat that would find out "on its own" and "quickly" what it needs to do to survive is precisely Rosenblatt's infinite perceptron. But for a living homeostat the entire world is the screen on which various shapes are displayed, and so its environment is extraordinarily complex. (It is a matter of complete chance which one of the infinite perceptrons learns to distinguish triangles from squares, which one the green of grass from the red of wild poppies, and which one self-preserving behavior from self-destructive behavior.) But it is enough if a single one of those perceptrons out of a quadrillion or sextillion survives—and with it the *principle* of its functioning—and we have the beginning of a regular, natural evolution understood as homeostasis. As we can see from this limited example (which does not attempt to re-create the whole biogenesis), all the "half," "quarter," and even "three-quarters" homeostats quickly die out in random environmental perturbations; only the truly "self-centered" and universally surviving systems remain, thanks to the "learned" procedures of self-preservation. This is the essence of the principial gap between physical objects, which are neither value-producing nor subject to valuation, and homeostats, which are "doubly axiological"—doubly because they demonstrate values (in their selection behavior) and because they themselves are subject to valuation (as better or worse homeostats) according to the criteria for *self-preserving* instrumental efficiency.

Organisms that would have to learn from scratch a certain minimum of functions to survive cannot exist in nature. The biosphere learned this important lesson through its emergence. Species differ mostly in whether or not their information capacity is distributed throughout their bodies. All physiologically active tissues in insects are informationally almost invariant, which means that at birth their hemocytes and neural ganglia already "know what to do." In contrast, mammals are anisotropic in the distribution of this knowledge: a suckling's white blood cells are equally saturated with information as those of a philosopher, but a suckling's brain is practically a vacuum when compared with that of a philosopher (or any adult, for that matter). Yet all organisms are homeostats from the moment of their formation—even a suckling's ignorance is not so great that it would fail to distinguish milk from gasoline as a source of nutrition. In this sense the axiological minimum, identical with the homeostatic minimum, is built into every organism. The survival directive is preprogrammed in every element of the set that biology studies.

The wide popularity of the term feedback should not overshadow the fact that feedback coupling is the main pillar of a system's self-preservation and the dynamic determinant of learning processes—but principially only in fully matured, true homeostats, not in systems that are still on the probabilistic path toward homeostatic equilibrium. If a system is exposed to perturbations that occur periodically in a sequence of repetitions that are identical, like playing a gramophone record, that is, if the environment behaves according to a "predetermined pattern," a system without feedback, having only sensors that trigger unidirectional reactions, can function as a homeostat in this environment. The only requirement is that the environment sends a single signal marking the beginning of the series of perturbations, to which the system responds appropriately. Obviously such a "deductively invariant" environment is possible only in special circumstances, for example, in a computer, which is why natural homeostats—living organisms—must utilize corrective feedback mechanisms even when their behavioral program is totally determined by genes (e.g., insects). Every stage of the realized program must fit the concrete and a priori unforeseeable variations in the environmental conditions—both initial and boundary. The great progress that cybernetics has made in studying animals, and a lesser success with plants, which lack a nervous system, suggests—precisely because of this difference—that the stability of homeostasis cannot be reduced to the informationally maximized effectiveness of regulation, if by this effectiveness we mean the universalism of the typically "orientational" behavior manifested as the network of connections between the homeostat and its habitat. Plant systems utilize the strategy of a *slow, generalized response* to large classes of stimuli (which do not necessarily have to be specific), lacking any tactical generalizations, especially short-term ones. As with animals, the strategy is minimax, allowing the environment to change some parameters of the system while maintaining the values of parameters essential for survival within a range required for life.

One could call it a "fuzzy" or indeterminate axiology. Not much more can be said on this topic, since the plant variant of homeostasis has not been as extensively studied as the animal one. We do not even know why "mixed" strategies are not possible. (Theoretically a system alternating between being plant and animal could gain a significant survival benefit; but such a "hybridization of principles" may not be possible construction-wise.)

In summary, there are two kinds of values. The first, instrumental, defines the degree of suitability of specific means to achieve a given goal, which, once achieved, may become the means to achieve the next goal, and so on. At the apex of this pyramid there is usually a value of the second kind. The value of bridge construction is instrumental, and the usefulness of a bridge can also be measured instrumentally; but if, observing that people travel on it, we ask why they do not stay at home instead, we get into an endless circle, because one instrumentalism leads to another, unless we accept an answer based on a value of the second kind: namely, that people do various things for survival, which is a value that cannot be reduced to any other. When we transpose this reasoning to the terrain of biology, instrumental values correspond to the measurable ability of systems to pursue the goal of survival, an ability embodied in their physical structure. One could of course argue that the acceptance of life by humans as the supreme value is also part of the human structure. These are just two formulations, in different languages, of the same thing.

A value of the second kind cannot exist if there is no subject to recognize it. (The recognition does not have to be conscious; it could be just a system of reactions marked with regularity in behavior.) In contrast, instrumental values are, one might say, objective, in that they directly depend on the features of the material world. The value of iron as a material for bridge construction existed before there were people on Earth. We do not create these values but only discover them in the world according to the scientific and technological knowledge that we have accumulated. This knowledge itself therefore has instrumental value because we do not create it by fiat but rather "extract" it from nature. But can we really claim that instrumental values exist in the absence of any homeostat?

It might be better to say that they do not exist if there are no goals pursued by anyone or anything, since these values usually mark an extent to which a given means is suitable for achieving a specified state. It is not necessarily incorrect to claim that the value of iron as a bridge material was established by the physical properties of the element before humans appeared on the planet, but we should realize that such a statement can easily lead to equating the existence of an instrumental value with the existence of a material object itself, which is false. Concrete conditions must be present for an axiological measurement to make empirical sense (to be intersubjectively verifiable, like any experiment). Instrumental value is the suitability to achieve a goal, and the more precisely we determine the goal that the valuated object or procedure is expected to serve, the easier it is to measure it. Some products of technology, like a bridge or a camera, have their goal built in. "Monotelic" systems are usually easier to valuate than "polytelic" ones: the more goals an object can serve, the more difficult it is to establish its instrumental value without any goal-related qualifications. If an object appears to serve no goal at all, its instrumental value is either immeasurable, that is, approaching the infinite (and then it becomes a value of the second kind), or it is zero. A star has no instrumental value for most of us, serving no purpose, but for a sailor it has the value of an orienting signpost. A human being, who has no "built-in goals" and can "do anything," that is, create any number of arbitrary goals for himself, also is not subject to instrumental axiology. So if I say that iron has an instrumental value, I implicitly place that element into the sphere of human technological practice.

"To be an instrumental value" is a relationship, as is "to be a disease agent." Such terms owe their measurability and general sense to the second, silent part of the relationship: value—with respect to what? the ability to cause a disease—in whom? Humans adopted iron for their purposes, but its properties that prompted them to do that existed before humans arose. Bacteria adapted to higher organisms, making them ill, but most of the features that enable bacteria to do this existed before there were higher organisms. So the more precisely we define the goal, the more exactly we can determine the instrumental value of anything that serves to achieve that goal—within the scope of a given technology. If we do not specify the scope, we may find in practice that a given goal can be achieved by infinitely many techniques. For this reason, instrumental values are always system-dependent. The instrumental value of food can be expressed with respect to everything that lives; the instrumental value of meat—to every carnivore.

Values of the second kind exist only when someone recognizes them. The consumption of goat meat elicits shock only if a person accepts the taboo against it, that is, if that person is properly informed. If the effects of what is recognized are incorporated into the system of adaptive functions of an organism and hence increase the chances for its survival (as a species or a person), then the value becomes instrumental. But when these effects are *superfluous* for all homeostatic functions—and at the same time form a nonrandom set—the organism manifests in its behavior (through its preferences) the presence of a value of the second kind. Values of both kinds can mix in behavior such that making a clear distinction may be practically impossible. (It is not necessary to know what the organism experiences subjectively; we treat it as a black box and search only for correlations between the sum of stimuli at the input and the sum of responses at the output.) Values of the second kind are either revealed through reasoning based on calculation (the sum of information usable for adaptation and information that is adaptationally useless but still systematically favored by the organism, both entering the global balance of the informational-transformative work of the system) or based on our systematic ability to recognize (without which we would be unable to discover the link between the death of the tribesman who ate the unspoiled meat and the taboo value of the latter).

The last general issue to consider is whether a homeostat itself is an object with a value of the first or second kind. This question is not about the homeostat's value-generating power, which it exercises by the choices it makes. When we declare X a homeostat, we are saying that X favors certain values and therefore lacks the indifference that characterizes a lump of dead matter. But how do we know that X (as a homeostat) is an axiogenic, teleological thing while Y is not—if, say, X is a lizard or snail and Y a planet surrounded with an atmosphere or a music box with dancing figures on its top? As I have already said, our decision will depend on a series of investigations and not a single observation. The presence of three kinds of factors in X will indicate that it is a natural (living) homeostat.

First, there must be a set of parameters of X that stay within a certain interval for X to continue being X. Second, there must be a set of parameters of X such that, when they reach certain values, X "ceases" to exist, but it creates, either at the same time or earlier, "offspring Xs." This second set has higher cardinality than the first.[7] And finally, for the whole class of Xs (for the whole species of organisms) there must be parameters, of still higher cardinality, such that when certain values are reached, all Xs "cease" to exist, together with their offspring, but their ceasing occurs through a transformation into Z, a different species, which means that speciation occurred. This is how new species of homeostats come into being.

According to modern biology, the set of Xs that we are studying now emerged from a sequence of previous sets, and there was a set of Bs or As that had no parental set but constituted itself from dead matter through a process of self-organization. Thus self-creation, self-preservation, self-reproduction, and the ability to evolve are the four characteristics of natural homeostats. When we model them, we do not have to include all four at the same time; these characteristics go together biologically but can be separated technologically. It is possible to build an apparatus capable of self-preservation but unable to reproduce, an apparatus capable of evolution but incapable of self-creation, and, possibly, an apparatus capable of self-creation but incapable of stable homeostasis. (If an object does not exhibit some of these characteristics, we may be confused, not knowing if we are dealing with a true homeostat or only an imitation of one, for example, a wind-up doll.)

All products of our technology are equipped with self-preserving characteristics. Some of those features may be an unavoidable component in their construction, but often they are designed to protect a device and thus serve its purpose only indirectly. Even a device constructed to destroy itself, such as a self-guided missile, must protect itself—*before* it reaches its target. The idea of a machine as a tool that not only performs specific tasks that serve a purpose external to it but also preserves itself leads to the idea of a machine that has no external purpose but only functions to maintain its existence for as long as possible. A machine is an object that carries out defined transformations, and since a given class of transformations determines the machine's purpose (if we know this class, we know the machine's purpose), the machine cannot be a subject of the transformations it carries out—unless it has a single use and its purpose is to destroy itself along with its surroundings, as with a bomb. For obvious reasons, no one builds machines whose sole purpose is to destroy themselves. (According to some interpretations of the second law of thermodynamics, the world is such a machine.) The idea of a machine implies an invariant group of parameters, and what determines the use and the purpose of a machine must lie outside that group. Because the preservation of a device can be treated separately from its purpose, which is documented by the existence of specialized disciplines (e.g., the study of material strength), technology can provide criteria for measuring the performance of a homeostat as a "machine for nothing except self-preservation."

Empirical measurability notwithstanding, we face a vicious circle: an instrumental value is the suitability of a means to achieve a goal, but the goal is to achieve the highest possible instrumental value. If we consider the parts of a natural homeostat separately, the purpose of the whole will seem predetermined. We can explain exactly and rationally what purpose is served by the organs of locomotion—for example, to gather food. Digesting that food is the purpose of another organ of the system, and we can examine how effectively this organ transforms food into energy which is then supplied, among other things, to the organs of locomotion. The circle closes: the organism moves to find food, which enables it to move. The engineer, although he provides the biologist with the scale and tools to measure the efficiency of each organ or even its overall fitness (given by the program of homeostasis), cannot—being an engineer—accept that such a device is rational. This troublesome circularity can be brought to the next level: if each generation serves the next, what is "the ultimate purpose"? We follow the food chain, plants serving animals, animals serving other animals, until we reach the level of the biosphere, of which we are certain that it serves no purpose and about which we can say nothing in an axiological sense. But much of this difficulty stems from the way we describe systems at various levels that we consider principally separate, which they are not. Because we focus on self-preservation at the level of an individual, then at the level of a species, then at the level of global evolution, organisms sometimes appear to be "autotelic mechanisms," sometimes appear to serve an external purpose, and what's more, various purposes in their teleology can be graded. Grass serves herbivores but also produces oxygen for all animals.

Let us avoid falling into the logical error of equating the sentence "This is a basket of big apples" with the sentence "This is a big basket of apples." That is, let us not go from the sentence "This is a set of purposeful devices" to "This is a purposeful set of devices." Individual devices may behave purposefully, but it does not follow that the whole biospheric set of them is purposeful. The search for the "purpose of evolution" is devoid of empirical sense if it goes beyond attempts to find the limiting distributions of the bioevolutionary process in time and space, that is, to find how the planetary development of life satisfies the conditions of ergodic theorems. We can, for example, treat individual systems, individual organs, as predictive devices, but we cannot treat the Earth's biosphere as a predictor. The integral behavior of the biosphere does manifest functions that can be identified as homeostatic or predictive, but as a homeostat or predictor it crucially differs from its organisms.

If in the future an active focus on values becomes equivalent to the optimization of ultrastable states, biology will meet physics somewhere in the middle. In so doing, biology will get rid of the ballast of the axiological notions that have grown anachronistic in their metainstrumental reach, and physics will include the instrumentalized issue of values into the theory of antientropic systems, a part of the general theory of physical systems.

II. Biology and Engineering

In what follows I introduce two basic ideas, that of a minimal homeostat and that of an ideal homeostat. A minimal homeostat is the realization of a self-preserving system that is reliable and at the same time energetically and materially the most economical in an environment with a specific amplitude of perturbation. An ideal homeostat exhibits the maximum self-preservation in the maximum number of perturbationally different environments. These are not definitions that a biological axiometrist would find useful, since they do not specify the many conditions that must be met. For example, "self-preservation" can be interpreted in a structural-material way, in just a structural way, or in a way that includes only those parameters that are invariant both materially and structurally.

Homeostasis, as behavior that resists perturbation, denotes a situation of conflict, which is why it can be described in the language of game theory. But this language may be misused. For example, pebbles rolling from scree into a stream might be treated as a kind of "evolutionary game." A pebble that has acquired a round shape experiences less friction when rolling and therefore goes farther into the water than others. The winner in such "topological evolution" is the pebble whose shape is closest to a sphere; the "goal" of the game is to get the round pebbles to the sea. This description may reflect the facts, but giving it a teleological dimension lacks validity, because the boundaries of the "system" and the events in it were defined arbitrarily (at variance with Ockham's razor: the facts do not provide sufficient justification for calling the pebbles in the stream an "evolving system").

If an axiometry of homeostats is permitted, there will exist very good, good, mediocre, and worthless homeostats, where the latter are homeostats no longer, just as an ax is not an object that can float.

I have been describing the value of homeostats in terms of self-preservation, an ability that is determined by a fitness test. The propensity of a homeostat to be valuated (axiometric) is coupled with its function of value-creation (axiogenic). In an ideal homeostat, this duality ("being valued" and "valuing") merges into one. When any perturbation in any environment can be coped with, no event is "harmful" and therefore "bad." For an ideal homeostat, it does not matter where it finds itself, because no external influence can either disturb or improve its internal state. The further a homeostat is from that end of the spectrum, the more axiogenic it is, because it shows "interests": events that have a negative value it will avoid and events that have a positive value it will pursue. At the opposite end of the spectrum, where homeostasis ceases, values disappear again, this time because for a thing unable to make use of any event for its own benefit, all events are equally worthless. This is the natural state of a pebble or handful of dirt: "inertness," "axiological neutrality."

Living systems, being continually sifted by the filter of evolution, are "good," that is, efficient homeostats. They are not ideal, because we can find perturbations that will destroy them. Devices that we build—computers, perceptrons—simulate some parameters of homeostasis but not its operational whole, which confounds us when we want to determine whether or not they produce values (by their choices).

With his constructions, the modern engineer fills the gap between a passive physical object and fully functional organic homeostat. But this gap was not devoid of entities even before there were people: it is hard to say whether a virus makes a decision to invade a cell or its action is merely a result of a specific reaction of chemical catalysis. The cyberneticists who build logical decision-making systems might argue that a virus does make a decision, but then chlorine molecules must do that too when they join with sodium to form table salt, and an ordinary doorbell will be a logical device that oscillates between the decisions "yes" and "no." The logicians are right to relate certain physical phenomena to the realm of logic; the selective isomorphism of physics and logic permits that. Only thanks to a similar kind of relationship, a perceptron constructor can assert that an apparatus, by recognizing certain shapes, assigns to them a positive value—and a negative value to other shapes. Such an assertion is allowed as long as we remember that we are dealing with *derivatives* of phenomena that were extracted from homeostasis but can reach full expression only within its boundaries. A constructor might eventually succeed in creating a true homeostat. He would have to choose between making a system that uses solar energy for self-preservation (the apparatus could then be stationary, because sunlight is everywhere on Earth, albeit periodically) or making a system that seeks sources of energy by moving around. Without any reference to the dichotomy of organic forms on our planet, just by pure reasoning, the constructor arrives at two alternative projects, functional homologues of plants and animals.

Validation of the teleological and, consequently, axiological approach in biology as an empirical discipline points to a research paradigm, but the modern researcher may not yet have the theory and technology to realize the program of objective axiological analysis in practice. It may not be an exaggeration to say that if biology had any expectation of help from other disciplines, engineering was at the bottom of the list. At least that was the situation forty years ago. Today it is precisely the aid of engineering that is responsible for success of biology, including its theoretical branch. The differences between an engineer and a biologist are obvious. A biologist studies systems that are *given*, that he did not construct, so he knows neither their "goal," in the functional sense, nor particular characteristics of their subsystems. An engineer always has a goal that guides him in his projects when he makes consecutive variants of devices, and yet—and I stress this—his predictive knowledge about the behavior of his products is usually incomplete. It is because an engineer, especially when a project is complex, works less on the basis of a theory and more by successive approximation, by testing a series of prototypes. This is the fundamental difference between the two approaches. Another difference between a biologist and an engineer is that each field, like every developed discipline, has its own shell of paradigms and rules, its own "cultural norm," as it were, which defines the limits of what is no longer in dispute, within which specific instrumental values may be determined. The point is that in engineering there is nothing that tells us whether the safety coefficient should be two or three times the normal (predicted) operational load of a structure. The rules in this technological shell, which cannot be derived from the shell itself, are known to every expert, since they are used in everyday practice. An engineer therefore knows how to balance economy in making a particular product with the redundancy in its safety reserves. A biologist rarely possesses such knowledge. What's worse, the leading discipline in natural sciences, physics, does not even recognize those categories, since atoms have nothing to do with either economy or safety. In this respect, then, the objects of biological research may be more similar to the products of an engineer than the bodies of a physicist.

Although engineering technology is slowly becoming a resource for biology in research methods, modeling, and formal structures, the devices it produces are still inadequate to put the biologist, who studies organisms, on an equal footing with the engineer, who studies machines. The formal analytical apparatus that biologists adopt from the discipline of techniques that can be called "cybernetic" is too simple for the complexity of biological systems. Not surprisingly, the apparatus has already become too simple for the engineers themselves.

Engineers cannot always tell—let alone measure and confirm—how close a construction is to the highest perfection that can be accomplished in a given branch of technology. Because of the progress in all branches, the bar is set higher and higher. The best combustion engine in 1940 is no longer the best in 1968; the best computer in 1949 cannot compete with the best in 1970. Yet this dual stream of progress—improvements of prototypes at a *given time*, a synchronic cross-section of technology, and improvements *throughout the path* that a particular technology spans diachronically from its birth to its demise (for example, large-sail ships are now extinct)—which compels instrumental axiometrists to constantly modify their criteria, does not lead to chaos. At any *given time*, the instrumental criteria that make instrumental axiometry possible are well defined. That the engineer does not know at the birth of a new technology the whole domain of its *theory* (which is the only thing that would allow him to make predictions both technological and axiometrical) is not a problem, because he learns as he goes, by trial and error; in the creation of the next generations of a device his knowledge increases as the device improves, thanks to the feedback between creation and creator.

Nothing like this exists in biology. The biology of the "simplest" organisms (e.g., some ciliates) is not theoretically or formally simple at all, nor is there a gradual transition from "elementary" organisms to those that are more complex, because even the most elementary are several orders of magnitude more complex than the theoretical models that we apply to them. For example, the regulatory systems of all organisms are nonlinear, not to spite the researcher but because this type of regulator is more effective, and there is little chance of finding an algorithm for a nonlinear system. We must resort to approximations, crude simplifications (e.g., by assuming that only one steering block is nonlinear while the rest are linear, which may help but is not true), or numerical simulations (which seem to work the best). This suggests that our knowledge is still incredibly primitive when compared with the information that living organisms have accumulated, starting with bacteria. How, then, is a biologist to practice instrumental axiometry instead of speaking about it in generalities? The scientist's handicap is best illustrated with an example. If an energy engineer from the mid-nineteenth century was asked to determine the instrumental value of a modern nuclear power plant, he would not be able to do it. Some subsystems of the plant might look to him quite strange but operational principle of some others would be completely beyond his grasp. If one knows neither the fundamental principles of a technology nor its limits (which requires a mastery of its theoretical basis), one can make no judgment about the efficiency or the degree of perfection of a given construction. If the nineteenth-century engineer were to apply the safety criteria established for a steam boiler to a nuclear reactor, I doubt that anything sensible would come of it. Preventing an explosion because of high pressure has little to do with preventing a nuclear reaction from running out of control.

That a biologist is in an even worse situation with respect to the objects he studies results in very different axiomatics of valuation in different branches of biology. Works by the "synchronists," researchers studying organisms presently living, take as a given the optimality of biological solutions. Insects may look primitive to a human anatomist but not to an entomologist. This is not meant as a criticism of construction premises in different types of organismal organization: we cannot criticize what we do not sufficiently understand. For example, we cannot question the need to sleep that all "cephalized" animals have, because we do not know what its purpose is. Works by "diachronists," that is, evolutionists, carrying out paleontological comparisons, argue that some forms were constructed "worse," that is, less effective anatomically and physiologically, than others. But it is easy to notice that these arguments tend to follow the model *post hoc, ergo propter hoc*: first they show that a species with a certain construction died out and then they feel compelled to seek—by speculation—the "construction flaw" that caused the extinction. Few refuse to advance such hypotheses and admit that looking for the cause of extinction within an organism itself is problematic. If a supernova wiped out the dinosaurs of the Mesozoic, no "construction flaw" was responsible.

These two lines of reasoning contradict each other. There were attempts to make them compatible, with the argument that better and worse organisms existed in the past but that evolution has come to an end and there never can be anything better than what we have now. This argument, popular in its time, has hardly any supporters today. The process of evolution continues (albeit with the perturbation of human culture superimposed on it). Even today, some species are being filtered out and others are thriving. How, then, do we make compatible "synchronic" biology, which only praises, with "diachronic" biology, which only finds fault?

The diachronic and synchronic can be combined only in this way: each consecutive stage of a species is perfectly adapted to its environment, but changes in the environment produce evolutionary gradients; what used to be the optimal solution to the adaptation problem ceases to be such, and the species adjusts. The total of taxonomic hierarchies therefore represents something like a vast tracking system that follows environmental parameters, and the coupling is regulatory. Yet this model is evidently wrong. As Stafford Beer noted in his book *Cybernetics and Management*,[8] the first vertebrates began flying not because their environment changed to air, even if earlier, fish gave rise to amphibians because areas of water dried out. But reducing the rise of amphibians to that cause is just as weak a hypothesis. The idea that expansion into new niches can be explained by "following environmental changes" is naive. We must recognize that the generator of diversity in organismal forms, given by genotypic variability, is creative process, not just a passive following of changes. If we revise the evolutionary-selection schematic accordingly, "optimal adaptation" appears to be a simplification, as there exist various kinds of responses, at different levels, to an identical adaptation challenge. It is precisely the phenomenon of the constant mutational inventiveness that reveals that the finitistic[9] picture of the evolutionary process implied by the idea of natural selection acting as a mere stabilizer contradicts the basic tenet of bioevolution.

Let us, then, admit our ignorance and accept the possibility that among the life forms that appear to us to be equally well adapted there are some that are being filtered out because they are inadequate and some that are undergoing transformation—and we cannot distinguish between the two. It is impossible to do so by studying the diachronic order, summarizing and integrating the results of a long series of selections, because our lifespan is too short; moreover, two centuries of biology's existence are a blink of an eye when compared with the time scale of evolution, too short for noticing these kinds of changes. And it is equally impossible to make the distinction by studying the synchronic situation, because we do not know all the intrataxonomic links. In early aeronautics, the diachronic approach would correspond to a series of observations that would inform us about the history of the lighter-than-air flying machines; the synchronic approach would correspond to the knowledge of their kinematics, steerability, and reliability in comparison with those of airplanes. The diachronic approach dispenses with the knowledge of the "immanent" properties of the flying machines, satisfied with knowing how the development took place *in fact* (i.e., that the balloons, blimps, and dirigibles lost to the airplanes in that "battle for survival"). In the synchronic approach, we do not have to wait for the end of the race but make predictions based on our knowledge of those devices' "immanence," that is, theoretical comparisons of the flight characteristics of airships versus airplanes. Our position in biology is more or less like that of the nineteenth-century engineer who knows that steamboats have triumphed over sailboats but seeing the first zeppelins and fuselages, with no option to wait for the outcome of their competition and not knowing the flight theory of either, is unable to tell which is headed to a bright future and which is headed to extinction.

The difference between the engineer and the biologist is that the engineer faces two forms that are equally primitive and imperfect and does not know their future, while the biologist faces two forms that seem equally perfect but he does not know which of their structural-dynamic characteristics is the key for their survival in the future.

III. Evolution Punctuated and Gradual

The parallels between techno- and bioevolution have their limits. An axiometric analysis of natural homeostats can be carried out in at least two ways: considering that the conditions of the beginning of life were either necessary or random. If they were necessary, we must accept them, and then we can question only the *decisional* sequences of the evolutionary process, because after the beginning, the process acquired a distinctly random character. This can be seen in the fact that where many populations participated in the "evolutionary game with nature," for example, on large continents, placental mammals appeared, but where the number of "players" was much smaller, as in isolated Australia, "only" marsupials evolved. This shows that the winning strategy in the game with nature depends on, among other things, how many "playing partners" form a "coalition" to counteract the "mutational moves" of nature, represented by the habitat. The more partners there are, the higher the chance that one of them will draw from the mutational raffle box a rare gene configuration that is the "main prize" of the game.

The most general formulation of the question is whether all the genotypic configurations that ever existed on Earth have exhausted the set of optimal homeostatic constructions. This question could be called "the problem of missed opportunities," that is, the winning tickets in the evolutionary lottery that no one ever drew. Evolution is a learning process taking place in a cruel school, where the punishment for failing is death and the reward for passing is life. I use the term "cruel" not in the moral sense but to stress the extremity of its measures, as it is hard to imagine a stricter method of teaching. The evolutionary selection is then a method of trial and error, in which the penalty for an error is death and the reward for a successful trial is—death's deferral. One might say that this strictness compensates for the slowness of the process, which is by principle Markovian and therefore has no cumulative memory: the best mutational inventions got erased from the stream of evolutionary improvement if the species that carried them was eliminated from the game because it made a mistake. This is why the same invention was arduously assembled from the elementary gene combinations many times *from scratch*.

In this light especially, two phenomena in evolution appear particularly important and, at the same time, surprising. (When we are surprised in science, it simply means that we do not know the causal mechanism.) The first is the construction universality of the genetic code, which shows that the information system that arose at an early, unicellular stage of evolution later exhibited a plasticity that could meet the construction requirements of all the multicellular forms of plants and animals that we know. This fitness redundancy is surprising since it established itself hundreds of millions of years before its actual use in evolution. Because the genetic code, both its "lexicography" and syntax, was originally—for a long time, perhaps a few billion years—an information tool for making systems with the complexity of an amoeba but later proved capable of producing organisms with much higher complexity, such as insects and vertebrates. One would have expected instead an exhaustion of the combinatorial power, and the entrapment in an absorbing subset of forms not very different from those that invented the DNA code.[10] We cannot explain this universality, acquired so early, unless we accept that there are deep, nonaccidental connections between the genetic code and language, namely, that both are principially *open* information systems with a set-theory character and similar numbers of degrees of freedom, owing to which the huge dissimilarity of their material substrates appears to be inconsequential. It follows that the rise of the genetic code and natural language represent two particular cases of the evolution of informational dynamic structures. The opinion that the hereditary code is a form of language has not yet been generally accepted, and most of the scientists who express it do so metaphorically. But I do not think that this is a metaphor; I believe that structural-linguistic research may finally enable us to understand the supreme laws that govern the emergence of all (not only natural) languages. Only then, with newly gained understanding, we will lose the sense of awe that we have vis-à-vis the chromosomal phenomenon, which will thereby find its place in the general theory of informational systems.

Today we know absolutely nothing about the generator that produced the language of heredity except that it must have been an enormously complex apparatus. The elements of the genetic code are not similar to any specific technology, because all our technologies have historically been finite, closed, unexpandable systems. That is the reason why they always reach limits of their possibilities, and we make progress only through our ability to abandon an old technology and turn to a new one. For example, the technology of converting thermal energy into electrical energy will have to be abandoned at some point in time, because there exists an efficiency limit of heat engines that cannot be crossed. We will switch to a nuclear method of generating electricity without using heat exchangers or to the direct conversion of the energy of chemical bonds into the energy of electrical current. During such an industrial revolution, a massive amount of knowledge, both theoretical and practical, stored in various constructions (e.g., the steam engine), is simply thrown away. Had evolution encountered such limits, it would have come to a stop, incapable of a total reorganization in which some solutions are stockpiled and others abandoned. Evolution can proceed only in a continuous way. (It is not continuous at the "quantum" level of genes, but this graininess is completely concealed in phenotypes by the compensatory-regulatory activity of ontogenetic buffers.) If biology is always a result of a summation of small changes, engineering, especially at significant turning points, proceeds *in steps*. We might argue about the scale of the bioevolutionary changes, but they can never compare in magnitude with the replacing of the steam engine with the nuclear reactor, for example.

But this difference is superficial, and the comparison is not rigorous. The basic principle of technological innovation is sequential change in energy sources, building materials, production tools, and the means of regulating those tools. In evolution, the energy sources, building materials, production tools, along with their control, are the same today as they were at the beginning: the energetics, the substrate, and the regulation have remained unchanged. More, any significant change in them appears impossible. There is no gene reshuffling that could make a newly emerging organism abandon chemical energetics for another type (e.g., nuclear) or change its building material or the rules of its transformations. It is only within this invariant triad that we can ask whether a combination of genes is possible that would allow an organism to organize itself in a way that is nontrivially distinct from the set that has been realized by evolution.

The magnitude of our ignorance will make the answer grossly simplified. Crude calculations suggest that the assumption about the complete lack of directionality in mutations and in selection of their outcomes is not able to generate a set of structures that would compete with the organizational diversity that evolution has achieved. Besides, the currently accepted theory implies things that are strange if not plain wrong. For example, the number of people living today already equals the sum of all their ancestors, starting with *Paleopithecus*. If *Homo sapiens* arose from that predecessor because of a mutation, then today's human population should exhibit no less variability than its parent populations. Consequently, any day now we should expect the emergence of a form of *Homo* that would be at least as new as the Cro-Magnons were with respect to the Neanderthals. But this is not happening.

After the death of entelechy,[11] there were attempts to embody its remains in genes—by bestowing on them a kind of omnipotence, the chromosomally located responsibility for everything that happens with an organism's phenotype. Some believe that there are genes that determine the chance of getting cancer, and that natural mortality is due to evolution's failure to eliminate lethal genes, just carelessly moving them from one corner of life to another, from the reproductive phase to the postreproductive phase, that is, the period of age-related decline. But in an automobile the destructive effects of its material's fatigue are not caused by an equivalent of a lethal gene: the fatigue is not a mistake of the project engineer. The logical model of embryogenesis, as transformations of original elements into their final organized set (determined by cell division), is a sequence of enumerable and ordered step operations. The logical depth of such transformations can be arbitrary, because deductive operations, whether informational or material, are error-proof.[12] But the logical depth of the embryogenetic process cannot be arbitrary, that is, there are construction limits given by the initial instruction of the genotype: after a certain level of complexity or a certain number of steps (the number of cell divisions in the embryo), the instruction loses its causative power and gradually drowns in "noise." The engineering concept that embryogenesis utilizes is that the total of causative information that creates the final product is ready at the beginning, in "a single packet"—the fertilized egg—and the process of creation needs no further regulatory "help down the road." In other words, a strong principle of engineering autarky applies, the same autarky that later gives trouble to doctors—helpers "unwanted by evolution"—who try to replace a damaged organ with a healthy one from another member of the species.

So lengthening the text of the original instruction and thereby increasing the depth of embryogenesis will not result in "completely new solutions" to the problem of homeostasis in the form of new organisms. A genotype, as a prognosis that realizes itself, has its causative *limits*. They were imposed on the genetic alphabet from the start and come in at least two kinds: limits "in width" and limits "in length." Limits in width are barriers that do not allow access to new energetics, materials, and regulation strategies; limits in length are barriers caused by "noise" that at some point exceeds the regulatory power of the structure. If genotypic engineering is helpless vis-à-vis limits in width, it may not be vis-à-vis limits in length: an embryonic process that because of its exceedingly high complexity begins to "trip on itself" can in principle obtain *"supplemental" regulation* from outside. In this way, genetic engineering could achieve states that natural evolution cannot. Such an achievement would be meaningless only if it turned out that beyond the space of constructions to which embryogenesis leads naturally there are no other solutions to the problem of homeostasis, that is, somatic regulators of the phenotype can stabilize only what "creative" regulators of the genotype create.

The combinatorial set of genotypes (artificial ones, made by the "genetic engineer") that can be constructed within such limits with the *given* DNA alphabet and its "syntax" has a cardinality significantly higher than the set of all the electrons in the universe. Even its subset whose members are only those reshuffling results that are true homeostats here on Earth has a cardinality that is not much less. However, a great majority of this subset's members will be trivial variations of actual organisms (such as a horse with a cloven hoof, for example).

The building blocks of all those constructions are the cells. As for their properties, we cannot expect any revolution there: despite appearances, undifferentiated cells do not differ that much from specialized ones—the parametric width is at most like that between the contraction of an amoeba and the contraction of a muscle or the stimulus transmission in an amoeba and in a neuron. These property differences are of course important for the organisms but quantitatively they have the same order of magnitude. A stimulus in an amoeba is transmitted with a speed of several tens of centimeters per second; in a muscle, between ten and twenty meters per second. Organisms on various evolutionary levels differ for structural-integral reasons, not because they are "squeezing" out of their building blocks some new abilities that were not present from the start. The physical parameters of the structure are given with an inviolable condition—that they are given once and for all.

The novelty level of homeostatic invention, then, is a function of the cardinality of all the configurations understood as structures created by the shuffling of genes. But this novelty, so limited in scope and variability, turns out to be questionable. With the gene alphabet it is not possible to build any energetics other than the current one, any locomotion apparatus other than the one that has been adopted (given the energetic restrictions) within the boundaries set by the skeletomuscular type of movement (the skeleton can be either internal, lever-axial, or external, plated, but nothing else), and any regulatory structures in information transmission other than those that exist.

With these limits and restrictions, at both micro and macro levels, what synthetic invention is still possible? Has evolution already realized all the options worth trying? That is doubtful: evolution "forgets," through the extinction of forms, various construction solutions, and we can "propose" them again to the genetic instruction. What's more, some radical systemic reorganizations seem to be possible—for example, replacing hemodynamics with another way of supplying electrons to the tissues instead of circulating oxygen.[13] It is doubtful that we can completely depart from the circulatory solution (i.e., pumping a liquid through a network in the body), but we might at least improve the pump. All animals use a pump, which is from the engineering point of view primitive (it is the *principle* of the pump that is primitive, not its realization; in an evolutionary aspect, this solution approaches the limit of what is possible, but the limit itself, given by physical properties of this type of pump, cannot be moved). Replacing a mechanical pump with an electrodynamic pump would not pose any serious problem, because, as we know, gene instructions are capable of making very competent electric organs. But equipping the moving corpuscules—the blood cells—with magnetic or electric polarity would be more difficult. No genotype can produce a magnet or, in general, a metallic part, and so will inevitably involve ions (no other solution seems feasible in the liquid environment). It might be necessary to concentrate ions above the tissue safety limit. So we can dispense with the localized pump that is the heart muscle and instead make the walls of all arteries an electrodynamic pump that has no moving parts.

We have come to understand not only that an electrodynamic pump is possible but also, what is more important, why evolution has not realized it. The prototype of the heart was a small contractible tube in small animals, and this solution was subsequently "dragged" through all branches of the evolutionary tree. It is a kind of solution that works the better, the smaller the animal in which it is "tried." For example, the tracheae in insects[14] make lungs and blood circulation superfluous but appear to limit body size, so insects could not "become smart": because the ratio of information-processing power to the capacity of the neural system is roughly constant, a neuronal brain cannot be miniaturized to such a degree that a moth or ant would reach the "intellectual level" of a rat. Had insects not stumbled on the tracheae solution, we would probably not be here. This is an example of how, on the evolutionary path, a solution to the problem of homeostasis becomes *irreversible*. Once the tracheae or the heart as a mechanical pump have arisen, it is impossible to back from them *through natural evolution*. Yet a transition from the discontinuous to continuous pumping of blood would bring advantages. Blood pressure would be stabilized, the problem of distributing the blood supply to different parts of the body would be simplified, and so on. The "instrumental axiometry" of homeostasis would definitely welcome such an "improvement."

Then why did evolution not "get this idea"? Regarding the opportunities that evolution "missed," I point to the possibility that a nontrivial innovation like this depends on the simultaneous (synchronous) occurrence of a number of mutations that are independent of one another. The higher that number must be for the "innovation" to take place in an organism, the lower the probability of the occurrence. Above a certain value, we could be speaking of an astronomically rare event, like tossing a coin a thousand times and getting a thousand heads. The evolutionary game would then be condemned to failure, were it not for a clever maneuver, "an ace up the sleeve," in the form of recessive alleles.[15] A recessive gene is something like a trump card: while "hidden," it has no power, but in certain situations of a game that is under way it can determine the outcome. A bridge player must wait a long time to receive a grand slam hand from a card distribution made random by shuffling. But with one trump card up his sleeve he can have a grand slam even if he is dealt all the necessary cards but one. The more trump cards he has hidden, the less time he must wait for the lucky hand. But just as a player cannot have a whole deck up his sleeve, an organism cannot keep in reserve an arbitrary number of recessive genes—especially since such genes are generally "not good for anything," that is, they do not find "instrumental values" in any of the possible population distributions. Besides, an organism cannot determine its own inventory of recessive genes, deciding which are "good to retain" and which are not. But the "chromosome engineer" will be able to do this someday.

Now we are getting to the point that is of special interest to the evolutionary axiometrist. Evolution is often "blamed" for the Markovian character of the regulator that governs speciation. Being Markovian, this regulator is uneconomical and extraordinarily slow in its learning. For this reason, many biologists considered that inheriting acquired traits—a non-Markovian type of chromosomal "learning"—was an evolutionary necessity. But evolution's method, for all its wastefulness, turns out to be better in the long run, because the equilibrium that a Markovian chain reaches is not final. The impracticality of the natural regulator, evident in the seemingly "meaningless" combinations that selection must constantly keep pruning away, is in fact a valuable source of variability, because only a high degree of variability guarantees the successful resisting of any change. Inheriting acquired traits is no doubt much more effective in the short run than a Markovian process, but it could easily lead the species into a dead end. A Markovian regulator allows the game to continue, although the price that evolution must pay for this freedom is huge: the wasted lives of billions of beings. I am not saying that this type of regulation is the best out of all that are evolutionarily attainable. A Markovian regulator may keep redealing the cards—in the "round" of amphibians, reptiles, and mammals—but its new decisions, governed by randomness, cannot be completely independent of the previous ones. The permanent supremacy of tactical solutions over long-term strategies results in limiting all future states, perhaps even billions of years from now, by the decision made for just one state in the present. In other words, the Markovian process is not immune to the possibility that future states that are homeostatically *superior* will be blocked by previous *inferior* solutions owing to the simple fact that those solutions occurred. A Markovian random generator cannot change the fact that the ancient amoeba had a much wider choice of evolutionary alternatives than the mammal. Evolution is thus a true game in the game-theory sense, in which good luck can beat any employed strategy and bad luck can bring an inescapable defeat.

The second way in which the evolution of life differs from the evolution of technology is the absence in the former of what is called moral obsolescence in the latter. Or it at least appears absent at first, as ancient corals coexist with the "modern" dolphin, the snail coexists with the human, and primitive lichens coexist with the latest botanical product of evolutionary inventiveness. This suggests that the set of evolutionary realizations is not ordered on a single axiometric axis. There are evolutionary tasks that may be solved in many engineering ways, but these solutions cannot be compared on the same scale. In reality, however, this is not different in engineering. In both biology and technology, problem-solving cannot avoid compromise, that is, the dilemmas that could be called engineering antinomies. These occur whenever a state that is optimal or maximal in one function represent, at the same time, a suboptimum or less than the maximum in another function that is equally important. An example is the relation between the deviation correcting and the tendency to oscillate, which always exists in an optimally regulated system. Such conflicts are usually more complex, having more parameters than two. It is an old wisdom that a whale cannot be as agile as a flea. Evolution works in an environment brimming with dilemmas like this.

I would add that an engineer's advantage over evolution is not as great as one might think on the basis that evolution does not have the luxury of prediction that an engineer has. The thing is that an engineer's information is incomplete and he cannot wait forever to achieve perfect knowledge about what he is constructing; in a way, every one of his inventions is "premature." The price to pay for decisions that are "premature" (in quotation marks because they cannot be other than that) is a high failure rate in technology (consider the first airplanes), and in evolution a high mortality rate (in a "novelty" radiation). A compromise is therefore unavoidable, whether the process is being carried out by a person or, as in evolution, "impersonally."

In any case the instrumental axiometry of biology must learn how to employ the methods that engineering has developed. Measures of value make sense only in the context of a particular technology: multidimensional comparisons can be drawn between a jet and a propeller plane, but one cannot compare a plane with a pair of roller skates. Intrasystem values of a particular technology have an equivalent in values found in a particular schematic of body organization in biology. Someday we might axiometrically order all insects or all land mammals, but we cannot ask (and even less so answer) whether the organization of insects is better than that of the mammals. The less particular and more general the thing that we are trying to measure is, the more obviously the measures of value lose their applicability. No axiometry can possibly exist that would objectively justify the claim that the human being is "the crowning achievement of creation." It may be an adaptation solution that is radically *different* from all others in the animal kingdom, but this "being different" can never become, in a purely instrumental sense, "being better." To receive this compliment we would have to supplement our criteria with noninstrumental values.

A "random-start" axiometry is mentioned here only briefly, because it goes beyond the limits of biology. The "inevitable" version, sketched above, accepts that the gene alphabet is a "given" and cannot be questioned. Yet we can imagine that all the forks and zigzags of the evolutionary process were not inevitable in every detail of every speciation or every event, and that the "lexicographic" and "syntactic" features of evolution's "articulation" apparatus—DNA—resulted from *random* events and therefore could have been *totally different*. This might be just a fantasy and nothing different could have occurred in reality, but until that is proven, a hypothesis of different contents is permissible.

As I have shown, the causative "articulation power" of the genes made from DNA is—from the technological point of view—not only far from being infinite, but it may not even surpass the capabilities of human engineering. For example, it cannot realize an energetics that is not chemical and does not rely on proteins. It cannot access processes whose realization requires high pressures, high temperatures, or high radiation densities. So the field of gene causality is clearly bounded and closed. We may still question the cosmic uniqueness of the initial alphabet and syntax. It is possible that different initial conditions, geological and chemical, could facilitate the emergence of a different "articulation apparatus," one not based, or at least not exclusively based, on proteins. If support for this is found either in a terrestrial laboratory or in the exploration of extraterrestrial bodies, biology will become a science that studies only one particular form of life processes. Which will obviously bring a new relativism into its axiometry, because what cannot be meaningfully valuated by an instrumental approach when it belongs to the set of genetically unrealizable constructions may be so valuated when it is created by another generator of heredity codes, offering us a broader configurational space for solutions to the problem of homeostasis. But today we can only think of this as a possibility because we know nothing of its concrete instantiation.

IV. Biology and Noninstrumental Values

Values "of the second kind" are typically a cultural phenomenon,[16*] which is well known to anthropologists, who dedicate all their efforts

---

*Naturally, with the exception of the value that life exhibits in its tendency toward self-preservation, this one is both pre- and metacultural. In this work, I am proposing an approach of "measured reductionism," which—following Ockham's razor and in accord with the

precisely to uncovering and comparing them. The question arises whether the presence of such values in biology can be detected in any objective way. I believe it can. But certain initial assumptions must be made, as in any kind of research. If in an organism any nonadaptive trait were immediately eliminated by natural selection, values of the second kind could not have emerged by evolution. If the environment had shown no "neutrality" with respect to at least some of such traits, it would be impossible to fit anthropogenesis into the evolutionary scheme.

Here too *ex nihilo nihil fit*.[17] Nonadaptive mechanisms, which in later stages of anthropogenesis manifested their axiogenic character (in the second-kind, noninstrumental sense), could not have arisen out of nothing, just as the human brain or a bird's wing could not have come out of nowhere: the predecessors of these organs must have already contained certain traits that were augmented in selection. It is therefore likely that not all traits in organisms are necessarily adaptive; some are additional, superfluous. These "add-ons" keep passing through the eye of the evolutionary needle because the environmental filter does not sift them out and because something facilitates their passage. Such facilitation can at first, in a small population, take the form of an ordinary "passive-statistical" genetic drift. Since the traits realized by the drift are generally not a substrate of any functional "meaning," they

---

empirical practice—tries to minimize the number of posited "entities," which manifests in the rejection of the axiological terminology based on ontological terms borrowed from instrumental techniques and theories. An "extreme reductionism" would deny the existence of any "rationally based," i.e., instrumental, values in biological phenomenology, and thus constitute an extreme "physicalization" of this scientific field, which would, however, result in throwing out the baby along with the bathwater, as the history of Watson's behaviorism documents—it proved to be untenable. Introducing axiologically pregnant terms into empirics was, until recently, incompatible with its tradition, i.e., with its research paradigm. At present, the terms are permitted to be used in nontraditional ways, e.g., in optimization schemes or dynamical programming. Still, a certain axiological minimum must remain unquestionable, i.e., accepted, that is to say, axiomatically. This means that some values (such as the "autotelic" value of being alive) cannot be reduced to anything else beyond themselves. The unavoidable property of anything that can be taken for instrumental value is that of to serve a purpose. Every irreducible value then implies a certain autonomous value. This proposed approach is equivalent to a convention, in particular, of accepting that there is a certain "homeostatic minimum," which can correspond to one on a comparison scale. Of course, we could accept another convention, but a convention of some kind is certainly necessary. The reason is that the overall paradigmatic structures of the studied range of phenomena cannot be directly deduced from the set of facts, and establishing them requires active input from researchers.

are instrumentally harmless but serve no purpose. If endorsed as a result of *sexual selection*, they might have participated as an "aesthetic" criterion in that selection. Discovering them is equivalent to discovering "autonomous values" in biology. Such hypotheses have been put forward, if in a slightly different form. Employing instrumental axiometry, we would conclude that a fancy courtship feather display in birds "is not good for anything" because a signalization promoting contact between the sexes can be achieved by much less showy and "aestheticized" feather patterns. If we definitively prove that the display is informationally redundant, that it does not combat environmental noise or augment the signal's species specificity, we must accept that it is the result only of the sexual partners' "aesthetic decisions." In that case a given feather pattern is preferred for reasons that go beyond signaling—the partner birds simply "like it" better.

Of course, biology is not permitted to speak in such terms; all it can do is note the nonadaptive redundancy of certain information. Most likely the reasons why that information has been privileged in sexual selection will remain a mystery forever. Only reasoning by analogy, which in a methodologically cumbersome and convoluted manner extrapolates from human to other organisms, allows us to ascribe this redundancy to esthetic criteria used by some animals (birds, lizards, etc.) for mating. But to determine what in that information is instrumentally superfluous, we must know in complete detail the utility boundaries of all information in an organism. Therefore, "the discovery of values of the second kind" must be preceded—and even that is no guarantee—by a thorough grasp of the instrumental axiometry of homeostasis. Unfortunately this goal lies in a distance of an unknown number of generations of biologists.

V. The Axiometry of Progress

The term "progress," when used in biology, denotes an increase in specific abilities. Not always, of course: we also say that a disease is "progressing" in an organism. But in general, progress means that the next state is in some way better than the previous one. The path that evolution has covered from amoeba to human being seems to be clear evidence that great progress was achieved. But when we subject this progress to axiometric tests, we encounter difficulties.

These difficulties do not appear when we compare the elements and functions of organisms in isolation. Different evolutionary solutions to the "problems" of seeing, hearing, blood circulation, locomotion, or formation of the "image of the world" in a nervous system can be ranked according to the level of their effectiveness. As it usually happens, if we have an aim specified in an articulated assignment, technological valuation is fairly straightforward. It might seem that to determine how far each solution is from the optimal point on our scale it is sufficient to transfer the results of our comparing to other forms of life plucked from the evolutionary tree. But this will quickly turn out to be nonsensical, because as the microscope is not the best watchmaker loupe or the lighthouse the best car headlights, so the eye of an eagle is not "better" than that of a fly, nor does a flatworm "improve" when given the organs of sight. The failure of this method prompts us to compare whole organisms as homeostats. But the best homeostat is not the biggest, not the one with the largest number of sensors, not the one that has the highest degree of harmony or complexity, and not the one that is thermodynamically the least probable. The best homeostat is simply the one that pursues self-preservation with the greatest success.

The task of homeostasis is not equally difficult in every environment; this difficulty depends on the level and quality of the environment's "noise." Ranking all the Earth's environments by difficulty is probably not possible, however, because a perturbation in one environment poses a *different* homeostatic task to solve in another. A good point of departure in our comparison would therefore be to introduce the notion of a homeostatic minimum for a particular environment. We could then proceed to create the notion of a "generalized minimum" as the ability to survive in two, three, or even more different environments (on land, in water, in air). Naturally, ecological classification distinguishes many local niches in each environment: adaptation in shallow waters is a completely different task from adaptation in oceanic depths. This fact notwithstanding, the notion of either a single-environment "minimal homeostat" or one "generalized" for several environments would be tenable if the course of evolution ran along either of those lines. But it doesn't: organisms with radically different structural blueprints can occupy the same ecological niche. Nor can the information theorem of heterogeneity (which says that a regulator must have heterogeneity sufficient to represent the environmental states) serve as a criterion. This theorem assumes that the regulation process—in our case, homeostasis—must act *continuously* and that the suspension of life functions is due to homeostatic failure. But such a suspension is often *reversible*. A homeostat, *subminimal* in a given environment, may not pit its heterogeneity against that of the environment but instead cease to be a homeostat temporarily. An engineer would appreciate such a trick and would love to be able to use it. When an airplane is about to crash, it would be wonderful if it could temporarily turn itself into a parachute. That this does not happen in reality is due to technical reasons, not any engineering principles. A person in an awful situation, instead of struggling to extricate himself, could hide in a freezer and in a state of "reversible death" wait for the situation to improve. Such behavior might be culturally judged as cowardice, but it is a value judgment distinct from instrumental valuation. As Scripture says, better a living dog than a dead lion.[18] Because no organism can withstand all perturbations, the ability to die reversibly in an emergency would benefit all living creatures. If evolution did not make this strategy universal, it must be due to enormous difficulties on the path to that solution, which modern medicine is trying to achieve in collaboration with biology. Artificial hypothermia has its equivalent in the natural hibernation of some mammals, but mammalian reversible death by "vitreous" freezing, of which biotechnologists dream, has no natural equivalent.[19]

Evolution creates "subminimal" homeostats, but it creates "redundant" homeostats as well, which are usually considered "progressive" forms. Evolution is a game that complicates itself over time. Its rules, initially limited to interaction between an organism and its environment, become enriched with rules for interaction between organisms, at first between organisms of the same species, then between organisms of different species. Locomotion solutions are simple when a predator ciliate chases a vegetarian ciliate but grow complex when a lion chases an antelope. Yet these two solutions are incomparable, and they remain incomparable if, for example, we replace the lion and antelope with a pike and a bream. All address the same homeostasis problem but for homeostases at different levels of redundancy.

Evolutionary progress may be real but is not rational from any technological point of view regarding homeostasis—because an engineer believes that a solution must be as simple as possible, that "constructed entities" should not be multiplied beyond what is needed. If we are in Europe and our destination is America, then any means of transportation that carries us from Europe to America faster than the one used before is a solution as rational as it is progressive. They are compared to the principially unattainable locomotion ideal, which is an immediate transfer, a transfer that takes no time at all and therefore marks an end on the scale of possible improvements. But if the task is the construction of a homeostat, there is no point in making the homeostat more complex. If an increase in complexity could bring us closer to the desired end of the scale, the ideal homeostat, and then the complexification, justified by the increasing homeostatic capabilities, would earn technology's approbation, whereby it would become subject to instrumental axiometry deducible from technology. But this is not the case. Higher organisms do not function better than lower organisms as homeostats, and an increase in complexity does not mean progress toward "a perfect homeostat." The difference between an ant and an antelope is like the difference between shooting craps and playing chess, not like the difference between a bicycle and an automobile. The latter difference is between two things that essentially have the same purpose but operate at different levels of complexity.

Evolutionary progress in organisms viewed globally amounts to the increasing investment of elements in a living system for the sake of better stabilization. The tasks that need to be solved become more complex in all aspects—in the material, energy, and both local and integral regulation—but our knowledge of the theories of regulation and dynamic programming, which is still negligible, suggests that although the difference between the information invested in an ant and an antelope may be physically measurable, it is not a sufficient basis for applying technological axiometry here. Neither an ant enlarged to the size of an antelope nor an antelope "microminiaturized" to an ant would be able to function. A hummingbird and an albatross may have the same construction blueprint, but they appear to be two different solutions to the problem of flight: an albatross is a long-distance glider, while the flight parameters of a hummingbird are more similar to those of larger insects. The baselessness of value judgments delivered when different organisms are compared causes the apologist of progress in evolution to prop his thesis—no doubt unintentionally—with ad hominem arguments. Julian Huxley, for example, compares the eagle to the flatworm and asks readers to picture the tremendous "progress" that has occurred between the two forms.[20] But what critical judging instance does this appeal address? Only our esthetic criteria make us imagine that an eagle's existence is wonderful and heroic, whereas that of a flatworm is opportunistic and ugly. These are not instrumental values, however. What's more, Huxley's words also imply an assumption that cannot be made in engineering. No engineer will say that the electrification of a city of one million is more progressive than the electrification of a hamlet, or that train dispatching in a large railroad network is more progressive than that in a small one. In each case, the former task is more difficult, but that's all. It is inconceivable that someone would ask us, "Which would you prefer to be, a small railroad network or a large one?" Yet it is possible to ask, "Which would you prefer to be, an eagle or a flatworm?" Empirically it is the same nonsense, but people will choose the eagle, not because "the eagle's situation" is "objectively better" than that of the flatworm—no one knows that—but because they feel that it is somehow existentially closer to them. The subtext of many an argument on the matter of evolutionary progress is, "The evolutionary higher an animal we are, the nicer (more appropriate, more interesting, experientially richer, etc.) it feels." We should not be surprised at the power of an argument like that.

A change in the environment means a change in the rules of the game. The factor that caused increasing cephalization of organisms or more generally their "individual ability to learn," was that the genotype's rate of learning could not keep up with the pace of environmental changes. Obviously the longer a personal life cycle, the slower the "chromosomal learning" will be, since the "quantum" of learning is the individual passage of an organism through the environment. In a simplified form the adaptation dilemma is: how can a regulator be built that performs well in the current version of the game but at the same time is prepared for any significant and *possibly rapid* change in the game's rules? Evolution has answered this question in a variety of ways. First, if the change is significant and rapid, withdraw from the game temporarily by "turning into a spore." But this tactic presupposes an eventual return of the original conditions (the game that the organism knows how to play). Second, maximize the preprogram heterogeneity of the regulator and make the preprogramming itself evolutionarily flexible. Finally, if possible, replace the preprogramming of the regulator with self-programming, learning capability.

In practice, the "answers," represented by organisms, are "mixed." The first belongs to the category of "spore" solutions, and it came first because reversible homeostasis is easier to realize the simpler the homeostat. The second answer makes up the class of solutions given by insects. Insects play an enormous number of preprogrammed games, with the preprogamming stabilized in the forms that passed through the environmental sieve many millions of times. The third answer, constituting the class of cephalizing solutions, is better than the second, but compromises between them are possible. The first solution retains a perfection of a kind that the other two have not achieved. The hibernation of some mammals is an attempt in that direction.

It is often said that insects are a lower form of animal that has shown a certain evolutionary success. Success and progress are not the same thing, however. Insects have existed for several hundred million years. Some build underground cities, others farm plants for food, and still others "domesticate" animals (other insects). Certain insects (bees) are the only animal form that has created a simple but effective language for instrumental communication that is hereditary. There are four times as many insect species as all the other species taken together. Insects are found in every environment on Earth except the oceans. None of the higher forms, which appeared considerably later, pushed the insects out of their ecological niches. In their time, the planet went through many random oscillations—mountains rose and fell; deserts turned into jungles and jungles into deserts; vast marshlands dried up, which finished off the saurians; ice ages froze once subtropical regions; the flora of whole continents changed—and insects passed through all those environmental sieves. When eventually humans appeared with insecticides, insects quickly adapted to them and their poisons. But this success is not an argument for the "progressiveness" of insects' construction. And if progress in construction is something other than the optimization of survival in multiple environments, there are no homeostatic criteria to measure it. Insects indeed fail to meet the requirement of "progressiveness," which according to Huxley is the power to evolve further: a progressive form should be not only optimally adapted but also able to change into a next, higher form.

It is easy to notice the problem with this definition, which implies a "duty" to change. In science, an analog of this evolutionary rule would be an empirical theory that not only predicts things but also is falsifiable, that is, it must be possible for data to disprove it. But if it turns out that a theory is always predictively effective, that data repeatedly confirm it instead of contradicting it, we accept after time that the theory is good. We do not think that it is the worse for not needing modification. Just the opposite: the larger area of facts it covers and the longer it lasts without change, the better it is.

One could say that these are a theory's "external" aspects, that is, in relation to the real world, the environment, and we valuate it accordingly. We can also valuate a theory in another way, by its internal properties: the criteria of structural simplicity, logical consistency, and so on. If someone were to show us a new theory that was equally successful in prediction as an old one but more complex internally, we would never say that the new theory represents progress for the reason that it was more clever in its structure or had more mathematical finesse.

Theories are *principially* valuated according to their predictive effectiveness; and what in science is predictive effectiveness, in evolution is the effectiveness of survival, fitness. Neither the "higher pleasures" of life that a larger brain may provide, nor the enormous complexity of the brain's structure, nor the regulatory ingeniousness of the body housing the brain justifies instrumental valuation of a mammal over an insect. Of course, we can employ another kind of valuation—by degree of complexity, for example. But then the higher value (the higher complexity) is a consequence of our assumptions, which are not *instrumental*. We would have to accept first that what is *more difficult* to make automatically merits *higher* valuation. But why should the probability of a state be inversely proportional to its value? No one knows.

Insects play the game against nature no less effectively than we do. A species that has survived several hundred million years does not need to "prove" anything, unlike a species that has been around for a mere six hundred thousand years. Insects passed the test; humans are just preparing to take it. No doubt insects are closer to the "homeostatic minimum" than we are. Their solution, simpler, has proved extraordinarily stable. Whether the cephalization solution is better, more progressive, is an open question worth considering.

The postulate of permanent plasticity is not absolute. It is relative with respect to the set of possible perturbations to the homeostatic process. If a form arises that can handle any perturbational eventuality, there is no "better" than that; there can only be a different form that solves essentially the same task in a different way. But is the task truly the same? Does not civilization introduce new conditions to be met, conditions that may contradict those created by the evolutionary process? When a civilization strives to minimize individual mortality, it exhibits a parallel with "engineering frugality" that contradicts evolution's approach, which does not minimize mortality at all. For evolution, preserving a species seems to be a task more important than that of preserving its individuals. And preserving the biospheric process itself is the most important task of all. Civilization attempts to reverse this hierarchy. But comparisons can be made only when the tasks are analogous. A flying species is not automatically better than a land one just because it can take to the air, and a species that conquers space is not automatically better because it can do what so far no other species has done. Organisms should be compared not according to a success of one particular solution but according to how they handle the "minimum task."

A recurrent theme of evolutionary progress is the thesis of the superiority of the cephalization solution over all others. The increase in neural mass noted in the diachrony of paleontological reports appears to confirm the universal benefit of having a brain and also the higher adaptivity of animals that have larger brains. Yet what exactly has a large brain, comparable in size to those of the primates, given to dolphins? A stable presence in the ecological niche of the shark, an undisputed "dimwit"—and not much else. In evolution, it is necessary "first to make the fish," then "turn it into the amphibian," and "through the reptiles" one can get to the mammals. Since a large part of the evolving organisms gets "trapped" on the way in the absorbing "sinks" of evolutionary immobilization, the solution that peaks in the neuronal approach—the anthropogenic—belongs to those that are the most difficult to achieve and are the least probable. But the maximum homeostatic value of this solution demands a separate justification, otherwise we are making a *circulus in explicando*: intelligence is the best, because the path to it is the longest, and the path to it is the longest, because intelligence is the best. That intelligence has created culture is a separate issue. We cannot valuate it in its immanence; we can only use the biological criterion of effectiveness. One of the two must obtain: either intelligence is the pinnacle of homeostatic self-preservation among all evolutionary solutions, or it is just one of many solutions, having no universal value.

Plants would survive without animals, but animals would not without plants. And this includes humans. At present, the maximum perturbation that the lower forms would survive in would be too much to handle for the higher forms. As spores, the lower forms could escape the destruction of a nuclear war, a nearby supernova, and other planetary or cosmic catastrophes to which civilization would succumb. Intelligence does enable us to note this vulnerability and thus make civilizational efforts to protect ourselves—but what measure do we use to mark the value of this diagnostic talent displayed by Pascal's reed? Especially when the optimization of evolutionary success always involves the overall equilibrium achieved by biocenosis. In the past there was so much talk about evolution's "tooth and claw" that people forgot that a predator that hunts too effectively, eliminating all its prey, dies out. Consider the evolution of parasites: the evolutionarily young parasites can be distinguished precisely by the fact that they are "too effective," killing their hosts and so endangering themselves. The evolutionarily old parasites work out a nonsuicidal equilibrium with their hosts. Evolution eliminates forms that are too exclusive in their self-preservation, making their narrow success temporary. Hence the form that defeats all other forms in the competition disturbs the ecological balance and the form turns into a self-destructive homeostat instead of the perfect one. Some "evolutionarily experienced" parasites show so much "caring" for their host that when there are too many in one host, they curb their "exploitation." Yet what in a long temporal series appears to be, in a purely physical sense, an inevitable outcome (an equilibrium that is simultaneously a criterion for the singular homeostatic effectiveness of the species and a dynamic state capable of further "complicating random walk" only when the stability of the entire system is preserved) is in reality a vast regulatory automatism which our so-called intelligent activities undermine instead of supporting it. Saving the biosphere is in our best interest, instrumentally, but we have not been very successful on this front. So civilizationally it is a long path from the emergence of intelligence as a homeostatic tool to making it safe, that is, eliminating its self-endangering potential. Intelligence may eventually become independent of the processes of the biosphere, at which point the biosphere will survive only if intelligence wants it to, but that decision will be based on humanitarian, not instrumental concerns. But of course, this refers to an unimaginably distant future, whose facultative state cannot be a measure of what is taking place now.

An opinion, sometimes formulated implicitly and sometimes explicitly in the context of the thesis of evolutionary progress, is that anthropogenesis is not a passing phase. The criterion for evolutionary progress, as Huxley clearly put it, is two-pronged: optimal adaptation in the present and the possibility for the evolutionary process to go "further" in the future. If the human being is the ultimate form, if there is no better adaptation tool than human intelligence, then we are a sink from which evolution cannot escape. But from purely organizational, statistical, and also adaptational points of view, machines that replace biological forms and create a global, autonomous planetary homeostat are a more robust solution than human civilization. Looking in the arsenal of evolutionary criteria for an argument that would make the value of civilization greater than that of planetary machine homeostasis is in vain: no such instrumental criteria exist. What to do then if some day cybernetic machines inform us that they are "the next stage of evolution"? If the principle of "transmissibility" as the determinant of progress is nonlocal, we must pack our bags. If it is local, we have articulated an apology for ourselves, for the system that created us. As we can see, granting intelligence the highest, ultimate value is a self-aggrandizing gesture that may turn out to be double-edged. We may lose the race of purely instrumental abilities, and therefore we should not consider the evolutionary-instrumental justification of culture as sufficient. Let us instead give civilization an autonomous value that is not deducible from anything. Of course, the machines that pretend to the position of our more perfect successors will consider that merely a desperate human trick—but this dispute with them is definitely beyond the limits of our topic here.

The strictness of my valuation of intelligence may be seen as disregard of values that have led to the "psychozoic culmination," but the rigor is just a consequence of the initial assumption—that of "falsifiable," experimental axiometry. From its point of view life is a self-supporting systemic response that is as indestructible as physically possible. This indestructibility is the homeostatic minimum environmentally bounded by the amplitude of biospheric conditions. In valuation, survival of individual homeostats, their personal longevity, apparently aiming for the principially unattainable goal of immortality, may be indistinguishable from the variant in which the system is stabilized by a frequent renewal (the "mayfly method"). Similarly, the expansive nature of life radiating from a single environment into all directions should be valued only by its resistance to perturbation; the expansiveness is not a value in itself, it is just the means to increase the number of tests for durability. As in engineering, the instrumental value of an object is determined by the result of the object's durability test, and the test in its technical aspects is subject to axiometry established by the measurement theory. The test, as a set of procedures, can thus be valuated regarding its ability to uncover the object's features that are of interest. In engineering, the test is relative with respect to the purpose that it serves; in biology, it is an end in itself, because it is given—by the self-preservation principle of homeostasis. If the products of evolution exhibit an increase in self-preservation effectiveness despite rising levels of noise, greater complexity can be explained instrumentally: for a given homeostatic minimum, the heterogeneity of the homeostat is proportional to that of the environment. But the overall bioevolutionary characteristics cannot be reduced to this proportionality, because it is only one of many determining factors.

Culture is an exceptional form of adaptation: first, because it is created by the other, metagenetic channel of homeostatic regulation (the channel that enables metachromosomal, cumulative learning), and second, because, unlike bioevolutionary change, a cultural change is principially *reversible*. This is suggested by the fact that the destruction of culture, whatever its cause, does not affect human biology; it merely pushes human beings back to the point of origin of sociogenesis (of course, within limits: a literally total annihilation of a culture is a historically rare phenomenon and the purely biological characteristic of the human being, the one that is solely determined by genetics and is completely peeled away from the layers of culture, can be—negatively—affected only by certain intentional procedures). Slightly digressing, I note that human efforts have usually tried to make a culture precisely an irreversible phenomenon and that the failure to achieve that goal has been considered a culture's weakness. But this phenomenon cannot be valuated according to typical evolutionary criteria. The rules of cultural development are not bioevolutionary rules, so bioevolution can teach us nothing about cultural norms and, vice versa, we cannot apply cultural criteria to evolution. Consequently, the point at which the bioevolutionary process leaves its traditional, monoselective, exclusively biological stochastics, that is, "the anthropogenetic locus of evolution," cannot constitute the pinnacle of the valuation scale that the biologist-axiometrist utilizes. The purpose that the properties of this point serve is no longer measurable on the scale of biological values. More, that scale itself changes at this point, and biology becomes valuated by culture. And this valuation is not necessarily instrumental. There is no contradiction between saying that "cultural criteria cannot be applied to evolution" and then saying that "biology is valuated by culture." The first, an "impossibility," simply means the absence of a place that is external to both culture and biology; the second, a "possibility," denotes the valuation of biology from inside a culture, which is therefore not completely empirical. Ascribing the greatest value to intelligence, stressing Schopenhauer's personally realized *"principium individuationis,"*[21] and striving for immortality, equally valuable as human strife for the "the ultimate truth" in epistemological determinations and for "understanding of the world"—all these are bound to be just *accidental* with respect to the evolutionary process. This accidental character applies to valuation, not to the facts that take place. Facts can be used to measure the degree to which particular features of evolutionary dynamics determine particular nuclei of human cultural endeavors (this influence is principally stochastic and therefore largely unpredictable). Neither these stochastics nor their consequences can be characterized as "good" or "bad" in a sense that goes beyond homeostasis. Even if all possible kinds of intelligence were equifinal states of many distinct planetary bioevolutions, the valuation of each bioevolution, made by the intelligent beings that are its own products, would be a declaration articulated from a random, arbitrary point of view. Even if the whole universe were bursting at the seams with intelligent beings, each and every valuation of the processes that produced such an outcome would remain instrumentally unjustifiable.

Life uses various tactics to realize homeostatic minima and homeostatic redundancies. We can call progress the increase in various abilities that takes place on the path along the branches of the evolutionary tree. But the scale that would allow us to measure how much instrumental goodness emerged between evolution's biogenetic beginning and its psychozoic end does not exist, because where there is not one path but an immeasurable many we cannot speak of values without ambiguity.

TRANSLATOR'S NOTES

1. *Studia Filozoficzne* (*Philosophical Studies*) was the most important philosophical journal in Poland between 1950 and 1990, whose Editor in Chief was the philosopher and the "historian of ideas" Leszek Kolakowski (1927–2009).
2. Axiology (from Greek ἀξία, "value, worth") is the philosophical study of value and valuation.

Dialogues

I

1. The Greek name Philonous means "lover of reason." In Berkeley's *Three Dialogues*, which served Lem as the matrix for his own *Dialogues*, Philonous personifies the author's views. The name Hylas is derived from the Ancient Greek *hyle* (ὕλη), which is "matter."
2. Kurt Lewin (1890–1947) was a German-American psychologist who is often considered the father of social psychology. He introduced genidentity (in 1922) as the existential relationship between consecutive moments of an object's existence.
3. *Ichor*—in medicine, the watery, acrid discharge from a wound or an ulcer, and in Greek mythology, the fluid circulating in the gods' veins instead of blood.

III

1. Lem is using the word "nebula" in its original, obsolete meaning, when it denoted basically any diffuse astronomical object, including the galaxies beyond the Milky Way. Canes Venatici is, in fact, a small constellation under the handle of the Big Dipper, visible to the naked eye. It includes several galaxies at a distance of about 20 million light-years.
2. Matter can be transformed into energy, but the opposite is also true. Einstein's famous formula $E = mc^2$ shows that matter and energy are equivalent.
3. For the sake of scientific rigor, we must note that the increase in disorder and the so-called disorganization of energy accompany only the sum total of all processes in the isolated system. If our isolated system contains a subsystem that is open, i.e., it can exchange matter and energy with its surroundings, processes may occur spontaneously in this subsystem, which increase its order—at the cost of increasing the disorder of the rest of our isolated system to a higher degree. And as for the term "disorganization of energy," modern physics uses the term "dissipation of energy." The "most dissipated" kind of energy is heat, which is characterized by a random, disorganized motion of particles in any direction and with any speed. Therefore, it is no accident that heat is a by-product of any transformation of one type of energy to another. Physicists then say that a part of energy during the transformation has dissipated into heat. According to Lord Kelvin

(1824–1907), the total dissipation of energy will lead to the so-called thermal death of the universe, when the latter will not contain any energy other than heat, which will be uniformly distributed throughout all of space. Then the universe will have achieved the final thermodynamic equilibrium, all processes will have ceased, and nothing can ever happen again. However, there are several reasons why the theory of the thermal death of the universe, based on equilibrium thermodynamics, is today considered obsolete. More about this can be found in the following pages of Lem's text and in the following notes.

4. Here, Lem means an isolated system composed of several bodies. If they do not all have the same temperature—thus, some are warmer than others—energy shows a certain distribution (order) in space. Uniform energy distribution, i.e., the lowest order, is reached when the temperatures of all bodies in the system are equal. The discerning reader will surely notice the connection with the thermal death theory, mentioned above in note 3.

IV

1. One of the prerequisites for the high velocity of a chemical reaction is the fluidity of the reactants, which allows for diffusion, and consequently, frequent encounters between the reacting particles. That is why it is difficult to envision the emergence of life in a stably ordered crystal in the solid phase; the alternatives are the fluids, i.e., gases and liquids. Gases are "too disordered": their particles, lacking any significant intermolecular forces, are in ceaseless chaotic motion. Only liquids, and particularly water, the "universal solvent," are a suitable environment for the chemical reactions of life. Nevertheless, in recent years, the view that surface chemistry may have played the key role in the emergence of life has drawn a lot of attention. The first organic reactions of life might have taken place at the boundary between the aqueous solution and the surface—either a crystal or layered clay. First, a decrease in the number of spatial dimensions from three to two significantly increases the frequency of collisions between the reacting particles and hence increases the reaction rate. Second, surface regularities (the clay's layered structure) may have acted as a catalyst (through arranging the reacting molecules into a suitable orientation) and a mold for the structural motifs in the molecules of life, many of which are polymers, that is, chains of regularly repeating units (the amino acids in proteins, the nitrogenous bases in nucleic acids, and simple sugars in polysaccharides).

V

1. This paragraph contains several assertions that require a critical reevaluation in light of the scientific and technological progress in the sixty-five years since the *Dialogues*' first publication.

First, it does not seem completely true that the upper limit of complexity, above which the system and its effective functioning start to break down, is only 100,000 elements. Currently the fourth largest and most powerful supercomputer in the world, Tianhe-2 at the National Supercomputer Center in Guangzhou, China, consists of 3,120,000 cores (https://www.top500.org/lists/2015/11/, accessed April 17, 2020), while each processor itself is a complex system of hundreds, if not thousands, of electronic equivalents of elements in Lem's "classical" understanding, such as relays, diodes, or transistors. However, Lem is correct in that supercomplex systems already cannot be homogeneous and usually they spontaneously "differentiate," both structurally and functionally, into more or less autonomous subsystems. In the case of supercomputers, it is precisely the core processors that function in the so-called massively parallel manner: the individual processors divide the whole complex task into partial subtasks and in a given time, each of them only works on its own subtask; the individual partial solutions are eventually integrated into the single, wholesome solution of the given task. In the case of "classical" living organisms, as, e.g., human, it is the individual organs that take responsibility for the individual partial functions of the integral life process.

As for the power (the energy per a unit time) consumption of supercomplex systems, Lem was not far from the truth. He just underestimated it a bit: Tianhe-2's power is 17.8 MW. For comparison, all power plants at Niagara Falls taken together produce a maximum of 4.9 MW, while utilizing more than 60 percent of the water from the falls (https://nyfalls.com/Niagara-falls/faq-4/#much, accessed April 17, 2020). The total possible output of Niagara Falls can thus be estimated as 8.5 MW, which allows us to conclude that despite the newest semiconductor and optoelectronic technology, the Chinese supercomputer requires for its cooling a power equivalent to double that of Niagara Falls.

2. This is from W. S. McCulloch and W. Pitts, "A Logical Calculus of the Ideas Immanent in Nervous Activity," *Bulletin of Mathematical Biophysics* 5 (1943): 115–133. Interestingly, several years later, other researchers discovered an error in the article, which, nevertheless had no significant impact on its main conclusion—see H. T. Epstein and A. Rapoport, "A Note on the McCulloch-Pitts Neural Net for Heat-Cold Discrimination," *Bulletin of Mathematical Biophysics* 13 (1951): 21–22.

3. In medicine, a state of less than full alertness, usually as a result of a medical condition or trauma.

4. Commissural fibers connect the left and right hemispheres in the brain.

5. Lem is using as an analogy here the description of now a long-obsolete TV system with a large, evacuated cathode tube, through which an electron beam was moving such that the electrons hitting the phosphor screen on the front face of the tube created a light image on it. Modern TV sets, of which no one in the 1950s could have had the slightest idea, do not use electron beams at all and work on totally different principles, which we are not going to discuss here because it is beside the author's point. Even the obsolete TV set principle is still a useful analogy for Lem's explanation of at least some operational principles of the brain.

6. The afferent nerves carry information from the sensory receptors to the central nervous system, while the efferent nerves carry the signals from the central nervous system to the effectory organs of the body.

7. See William Shakespeare, *Hamlet*, act 2, scene 2.

8. If Lem had in mind the number of the neuronal cells in the human brain, his number is too low. The current estimate for the number of neurons in man's brain is 85 billion (S. A. C. Azevedo, L. R. B. Carvalho, L. T. Grinberg, J. M. Farfel, R. E. L. Ferretti, R. E. P. Leite, et al., "Equal Numbers of Neuronal and Non-neuronal Cells Make the Human Brain an Isometrically Scaled-Up Primate Brain," *Journal of Comparative Neurology* 513 [2009]: 532–541).

VI

1. In Greek mythology, Atlas was a Titan god of endurance and astronomy. After the suppressed Titans' revolt against Zeus, Atlas was condemned to support by his shoulders the vault of the heavens for eternity. Interestingly—and illogically—in many Classical pieces of art, Atlas is seen holding up the Earth's globe instead, and apparently Lem also fell victim to this "fallacy."

2. Nicolas Rashevsky (1899–1972) was a Ukrainian American theoretical physicist, a pioneer of mathematical biology, and the founder of mathematical (theoretical) biophysics. In 1939, he founded the influential journal the *Bulletin of Mathematical Biophysics*. He appears to have contributed to the development of the first mathematical model of the neuronal network, which this journal published, by his student Walter Pitts with Warren McCulloch in 1943 (see note 2 in Dialogue V).

3. Rafael Lorente de Nó (1902–1990) was a Spanish American neuroscientist who significantly contributed to the understanding of the structure and function of the cerebral cortex.

4. The author is alluding to at that time very fresh "discovery" of the American psychiatrist R. G. Heath, which, after its publication (R. G. Heath, S. Mårtens, B. E. Leach, M. Cohen, and C. Angel, "Effect on Behavior in Humans with the Administration of Taraxein," *American Journal of Psychiatry* 114 [1957]: 14–24), achieved quite a fame. Unfortunately, Heath's results could never be replicated in other laboratories, and the case is therefore considered today a classic case of scientific self-deception, if not a scientific misconduct. Yet this doesn't mean that the material cause of schizophrenia is out of the question: the possible genetic, immunological, biochemical, and other molecular causes of this disease attract significant attention from biomedical researchers.

5. The British pioneer of computer science, a mathematician, logician, cryptoanalyst, and theoretical biologist Alan Turing (1912–1954) unveiled his universal automaton, also called Turing's universal machine, in 1936 and 1937. This "computer" is a purely mathematical, theoretical construct, which can demonstrably solve any solvable mathematical problem and, despite its simplicity, simulate any operations that are possible, including those performed by systems with arbitrarily high complexity, such as the human brain. The machine is utterly simple: it consists of just a reading and writing head, which can transition between a few defined states and can read and print a few defined symbols on the data tape that runs in front of it, and a shifting mechanism, which moves the data tape—according to the instructions on the tape and the current state of the head—one frame forward or back. The more complex the process that the universal machine is simulating, the longer the data tape and the longer the simulation. Thus for very complex processes, it may take almost an eternity for the automaton to arrive at the solution. The mathematically gifted reader with an adventurous character can check out Turing's original paper (A. M. Turing, "On Computable Numbers, with an Application to the Entscheidungsproblem," *Proceedings of the London Mathematical Society* 2, no. 42 [1937]: 230–265).

VII

1. This and the following sketch have inconsistent legends, perhaps owing to misunderstandings between the original editor and the author. In both cases, the stimulus is an ordinary, regular sinusoid curve, which appears to be missing from the first plot. Its solid line represents the sum of the stimulus and the reaction. Perhaps for the sake of clarity, the latter (dotted line) is plotted with the opposite sign. In the second plot, the full line is correctly displaying the regular stimulus and the dotted line the reaction, which is inevitably rising due to the positive feedback.

2. It is a common misnomer when the universal constant, the maximum possible speed in the Universe (300,000 km/s), derived from the theory of relativity, is called simply "the speed of light." Light only propagates at this speed *in the vacuum*; in material media, light travels with speeds that are lower (sometimes much lower) by a factor given by the

media's refraction index. Thus, the correct name of this constant is "the speed of light in the vacuum."

3. Vito Volterra (1860–1940) was an Italian mathematician and physicist, who contributed to the development of mathematical biology and functional analysis. The differential equations that Lem is mentioning are today called Lotka-Volterra equations, as the US mathematician and physical chemist Alfred J. Lotka formulated the same equations, independently from Volterra and in slightly different context, a few years earlier (A. J. Lotka, "Analytical Note on Certain Rhythmic Relations in Organic Systems," *Proceedings of National Academy of US* 6 [1920]: 410–415, https://www.ncbi.nlm.nih.gov/pmc/articles/PMC1084562/pdf/pnas01916-0016.pdf, and V. Volterra, "Variazioni e fluttuazioni del numero d'individui in specie animali conviventi," *Mem. Acad. Lincei Roma* 2 [1926]: 31–113).

4. It is not clear what the basis is for this claim and what new species could have triggered the megasaurians' extinction. Currently, the majority of the scientific community supports the theory of a sudden climatic change following the fall of a huge meteor, even though it does not completely account for all known facts (see, e.g., https://australianmuseum.net.au/the-mesozoic-extinction-event, accessed April 18, 2020).

5. The Englishman John Maynard Keynes (1888–1946), often regarded as the most influential economist of the twentieth century and the founder of modern macroeconomics, showed that the so-called free hand of the market alone cannot keep in check the oscillations in the socioeconomic system so as to secure the optimal flow of economics and minimize the negative impact of the increasing oscillation on the system's elements, i.e., people. He summarized his thoughts in the book *The General Theory of Employment, Interest and Money* (London: Palgrave Macmillan, 1936). Keynes defended the intervention of the state through fiscal and monetary instruments at its disposal as inevitable. With a little hyperbole, one might say that he attempted to give capitalism a "human face." While his theory has never been accepted without reservations and was sharply criticized mainly in the 1970s, the international financial crisis of 2007–2008 revived it, and the so-called new Keynesian economics is a widely accepted pradigm today.

6. John von Neumann (1903–1957), a Hungarian American polyhistor, made invaluable contributions to the development of many scientific disciplines such as mathematics, physics, economics, and scientific computing. He was the key contributor to concepts such as cellular automata, the universal constructor, and the digital computer. Here Lem alludes to the mathematical game theory of which he was the founding father (J. von Neumann, "Zur Theorie der Gessellschaftspiele," *Mathematische Annalen* 100 [1928]: 295–300, and a later book, J. von Neumann and O. Morgenstern, *Theory of Games and Economic Behavior* [Princeton, NJ: Princeton University Press, 1944]).

7. Lem is playing with the phrase from the work *Apologeticus* by the early Christian theologian and apologetic from Carthage, Tertullian (ca. 155–240), *testimonium animae naturaliter christianae* (*Apol.* 17.6), by which he meant that every soul has a Christian nature.

8. In Greek mythology, Procrustes was a demigod (allegedly the son of Poseidon, the god of the seas), a blacksmith and robber by trade, who appeared to be something of a jolly good fellow offering lodging to travelers. He had, however, an obsession: when he laid the traveler on his iron bed he could not find peace until he fixed the traveler's length to be precisely equal that of the bed. That means that if something was sticking out, he cut it off, and if something was missing, he stretched the unfortunate individual. This gave him the nickname Procrustes, which means "The Stretcher." Today, his name is used to denote the principle according to which something is forcibly adapted to an arbitrary standard with

no regard to costs, even when logic would dictate the opposite, i.e., to adapt the utilized standard to the natural existing conditions.

9. It would not be Lem if he did not attempt to teach us something on the side. This time it is Latin declension. *Generis humani* ("of human race" in English) is the singular genitive case of *genus humanum*. Coincidentally (or not?), the encyclical of Pope Pius XII, promulgated in 1950, dealing mostly with philosophy and theology, but extending also to culture and science, was entitled the same: *Humani generis*. Interestingly, it explicitly mentioned "the doctrine of evolution," with the pope taking a nonconfrontational attitude toward it "as far as it inquires into the origin of the human body as coming from pre-existent and living matter" and leaves the human soul in the exclusive jurisdiction of God. (The whole text can be found on the website of the Vatican Observatory Foundation, https://www.vofoundation.org/faith-and-science/humani-generis-1950/, accessed on June 21, 2020.)

VIII

1. Gerardus Mercator (1512–1594) was a Flemish cartographer who in 1596 published a map of the world in a novel, cylindrical projection. His map was especially welcomed by the seafarers because a straight course on the sea is represented by a straight line on the map, too, that is, the straight course crosses all meridians, which are parallel with each other, at the same angle. However, the Mercator projection has one drawback: it deforms the shape and size of large objects, and the more so the farther they are from the equator. The reason is that the meridians are truly parallel only at the equator and they all cross each other at the poles. Consequently, the poles in Mercator projection do not exist. Or more precisely, they do not exist as points; rather, they are represented by the whole length of the upper and lower border of the rectangular map.

2. Neurosis is a spectrum of mental disorder that causes a sense of distress and deficit in functioning but not incapacitation. Usually it has no discernible somatic causes. Examples include obsessive-compulsive disorders, anxiety, depression, and post-traumatic stress disorders. There is no universal consensual definition of neurosis. Modern psychiatry no longer recognizes it as a *bona fide* psychiatric illness, but that was not the case when Lem wrote the *Dialogues*. The term was removed from the *American Psychiatric Association's Diagnostic and Statistical Manual of Mental Disorders* in 1980.

Neurasthenia is a similarly obsolete term with an even more vague definition: its "symptoms" include chronic fatigue, inability to concentrate, insomnia, and loss of appetite, among others.

3. The German psychologist Ernst Kretschmer (1888–1964) was known mainly as founder of so-called typology, that is, the relationship between the personality and the body type. However, his classification scheme, which he attempted to apply also in psychopathology, finds no great support in modern psychology, to put it mildly. Lem refers here to Kretschmer's statement, which in 1933 he allegedly said to colleague Oswald Bumke, who cites it as follows: "It is a strange thing with psychopaths. In normal times, we judge them, but in the times of political upheaval, they rule over us" (O. Bumke, *Erinnerungen und Betrachtungen: Der Weg eines deutschen Psychiaters* [Munich: Richard Pflaum Verlag, 1925], 123).

4. The book *Facing the Extreme* (New York: Henry Holt, 1996; translation of the French *Face à l'extrême* [Paris: Editions de Seuil, 1991]) may be just the analysis that Lem was calling for. It was written by the Bulgarian French historian, philosopher, sociologist, literary critic, and essayist Tzvetan Todorov (1939–2017). In this work, he analyzes and generalizes experiences, testimonies, and opinions of the eyewitness survivors of the concentration camps and gulags.

5. Osculation (in mathematics) is a contact of two curves or surfaces at which they share a common tangent.

6. The first original Russian edition was published in 1864. The quotes in English are taken from the free e-book available at http://www.planetebook.com/ebooks/Notes-from-the-Underground.pdf, accessed April 18, 2020.

7. Just a few references to show how this topic is tackled today. The first is a quotation from the text of the US essayist, editor, and blogger David H. Freeman, "The War on Stupid People" (*Atlantic*, July/August 2016, https://www.theatlantic.com/magazine/archive/2016/07/the-war-on-stupid-people/485618/): "When Michael Young, a British sociologist, coined the term *meritocracy* in 1958, it was in a dystopian satire. At the time, the world he imagined, in which intelligence fully determined who thrived and who languished, was understood to be predatory, pathological, far-fetched. Today, however, we've almost finished installing such a system, and we have embraced the idea of a meritocracy with few reservations, even treating it as virtuous. That can't be right. Smart people should feel entitled to make the most of their gift. But they should not be permitted to reshape society so as to instate giftedness as a universal yardstick of human worth."

The second is an article by the Czech politologist and sociologist living in Germany, Petr Robejšek, titled "Why the Mass Is Smarter Than the Elite." It was originally published in the journal *Respekt*, but its expanded version is available on the website http://blog.aktualne.cz/blogy/petr-robejsek.php?itemid=27483, accessed April 18, 2020.

The third reference is to the the book *The Wisdom of Crowds* by the US reporter James M. Surowiecki (New York: Anchor Books, 2004). The book has a telling subtitle: *Why the Many Are Smarter Than the Few and How Collective Wisdom Shapes Business, Economies, Societies and Nations*. The author lists proofs of the thesis that if we define the correctness of a solution as a benefit to society and the majority of its members—which is not dissimilar to Lem's—we will find that the decision's correctness predominantly depends on the size and composition of the group that is making the decision and on the democratic discussion along with the plurality of opinions in it. In other words, larger groups that include lay people make better decisions than a handful of experts in a given field.

Supplement 1: The *Dialogues* Sixteen Years Later

Lost Illusions, or From Intellectronics to Informatics

1. The Austrian biologist Karl Ludwig von Bertalanffy (1901–1972) worked out his general systems theory in the 1940s and published its first wholesome account in English in 1950 (K. L. von Bertalanffy, "An Outline of General System Theory," *British Journal for the Philosophy of Science* 1 [1950]: 114–129). His metadisciplinary concept had an ambitious aim: to describe the behavior of a system consisting of interacting parts, regardless of whether the system is biological, cybernetic, or even societal. In 1956, the Society for General System Research was established. Yet the development of systems theory as an independent science was neither linear nor long. Today it can be considered a part of the so-called complex systems science, whose main interests are emergence and self-organization. Emergence can be understood as "self-creation" or "self-appearance" of qualitatively new, unexpected, and unpredictable properties when the system crosses a certain threshold of complexity; self-organization is one of the fundamental properties of highly complex systems. A number of other science disciplines, e.g., the network theory, pattern formation, collective phenomena (such as phase transitions), nonlinear dynamics, game theory,

and the theory of evolution and adaptation, can now also be considered parts of the complex systems theory. An excellent explanation of its fundamentals can be found in the book by the US physicist Yaneer Bar-Yam, *Dynamics of Complex Systems* (Reading, MA: Addison-Wesley, 1997). A higher-level college textbook, it can be safely recommended only to those who have working knowledge of physics and higher mathematics.

2 Algebra studies mathematical symbols and the rules of their manipulation. The word is derived from the Arabic *al-jabr* (الجبر), which in medicine denoted something like "putting back together broken parts." As the Arabic origin of its name indicates, algebra's founding fathers were Arab-Persian mathematicians, such as, Muhammad al-Khwarizmi (780–850), after whom we named the algorithm, and Omar Khayyam (1048–1131), whom people may know better for his poetry. Boolean algebra is a branch of algebra in which values of the variables are not numerical but "true" or "false." It is easy to label these two values 1 and 0, respectively, which later made Boolean algebra the basis for the development of digital computers and digital electronics. The English mathematician George Boole (1815–1864) presented his new algebra in his book *The Mathematical Analysis of Logic* (Cambridge: Macmillan, Barclay and Macmillan, 1847).

3. Claude E. Shannon, in his book *The Mathemathical Theory of Communication* (Urbana: University of Illinois Press, 1949), wrote: "[The] semantic aspects of communication are irrelevant to the engineering problem" (3). Although in some sense that may be true, many scientists just could not accept it. The first sign of research in this direction was a paper from Research Laboratory of Electronics at MIT in 1952 (R. Carnap and Y. Bar-Hillel, "An Outline of a Theory of Semantic Information," *Technical Report No. 247* [Cambridge, MA: Massachusetts Institute of Technology, 1952]).

4. Biocenotic population is an ensemble of various organisms and species that form an interconnected community.

5. I have not been able to trace the origin of Lem's numbers, but I am afraid that something is not right here. First, the thermodynamic unit of entropy is joule per kelvin (J/K). Based on the purely formal relationship between the informational, S_{inf} and thermodynamic entropy S_{TD}, which is $S_{inf} = S_{TD}/(k \ln 2)$, where k is the Boltzmann's constant, the rounded equivalence between the bit and the J/K is: 1 bit = 10^{-23} J/K or 1 J/K = 10^{23} bits. A machine consisting of 10^8 digital elements (where each can be only in one of two states) may contain at most 10^8 bits, which is equivalent to 10^{-15} J/K. As Lem correctly notes, this value is very, very small. But increasing the complexity of the system to 10^{12} elements is insufficient for any measurable change in the thermodynamic entropy, since it will be still a negligible 10^{-11} J/K. To support his argument, Lem would need a system with a complexity of 10^{23}, plus or minus a few orders of magnitude. Only in such an enormous system might the informatics and thermodynamics have comparable effects on its behavior. By the way, it is not without interest to notice that the number of functional molecules in the human brain has about the same order of magnitude.

6. L. Brillouin, *Science and Information Theory* (New York: Academic Press, 1962).

7. The British psychiatrist and the pioneer of cybernetics as well as the complex systems science W. Ross Ashby (1903–1972) mentioned the intelligence amplifier at the end of his book *Introduction to Cybernetics* (London: Chapman and Hall, 1956). The book is freely available for non-profit educational and research purposes on the Principia Cybernetica Web (http://pespmc1.vub.ac.be/books/IntroCyb.pdf, accessed on May 6, 2020).

8. The Ancient Greek word εὑρίσκω ("I learn or discover") has a Latin equvivalent in *inventio* and means finding a solution in an informal, nonalgorithmic way, often empirically (trial and error) and trading rigor for speed or efficiency.

9. Turing's universal machine has already been mentioned in Dialogue VI and commented on in note 5 of that section.
10. "God is subtle but he is not malicious." Einstein supposedly said this in April or May 1921 at Princeton University and it is often linked with his distaste toward the paradoxes of quantum mechanics.
11. "With joint effort" (Lat.).
12. A random, stochastic process is Markovian when it has no memory, that is, each of its consecutive steps is totally independent from any previous steps.
13. From the Greek αὐταρχία, which means "self-governance."
14. "Body" (from the ancient Greek σῶμα).
15. The "Danaian gift" is closely related to the more familiar term "Trojan horse." It is a gift in which future disastrous consequences for the gifted lurk. In Homer's epic *Ilias* or Virgil's *Aeneas* (II, 49), the inhabitants of Troy, which was located on the eastern tip of the Asia Minor peninsula, call the mainland Greeks Danaians.
16. "But who will watch the watchmen themselves?" Juvenal, *Satires VI*, lines 347–348.
17. M. Taube, *Computers and Common Sense: The Myth of Thinking Machines* (New York: Columbia University Press, 1961).
18. The reference is to the work of the Chinese-American mathematician, logician, and philosopher Hao Wang (1921–1995), "Toward Mechanical Mathematics," *IBM Journal of Research and Development* 4 (1960): 2–22.
19. Hubert L. Dreyfus (1929–2017) was a US philosopher and a critic of the artificial intelligence program, who rose to fame with his book *What Computers Can't Do: The Limits of Artificial Intelligence* (Cambridge, MA: MIT Press, 1972) and its updated version *What Computers Still Can't Do: A Critique of Artificial Reason* (Cambridge, MA: MIT Press, 1992). His criticism is based on the thesis that intelligence is not merely the ability to process information and manipulate symbols, which can be easily programmed, but rather an intuitive recognition of patterns and a subconscious ability to form unexpected connections between them, which cannot be captured by a logically constructed program. A logical consequence of this thesis, therefore, is the view that computers can be equal to humans only when they also mature under their parents' guidance in a complexly structured society—when they become human, like ourselves. Hubert's brother Stuart graduated with a degree in applied mathematics from Harvard University in 1964. Together with Hubert he authored several articles and the book *Mind over Machine: The Power of Human Intuition and Expertise in the Era of the Computer* (New York: Free Press, 1986).

Applied Cybernetics: An Example from Sociology

1. This is an allusion to the medieval treatise *Malleus Maleficarum* (1486), commonly known as *The Hammer of Witches*, attributed to the Dominican monk Jacobus Sprenger. The book describes contemporaneous beliefs about practices of alleged "witches" and may provide insight into medieval judicial proceedings, but later it instigated the savage efforts to stamp out "witchcraft" in Western Europe in the sixteenth and seventeenth centuries.
2. In Aristotelian logic, apodictic propositions are demonstrably, necessarily, and self-evidently always true. From the ancient Greek ἀποδεικτικός, "able to be demonstrated."
3. Jan Szczepański (1913–2004) was a Polish sociologist and politician. Lem's text does not allow for a precise determination of which text he is referring to, but a good summary of Sczepański's views is his essay "Individuality and Society," which he published in 1981,

a year before his retirement and the year in which martial law was declared in Poland (J. Sczepański, "Individuality and Society," *Impact of Science on Society* 31 [1981]: 461–466).

4. This folk fable from Central Europe may have a distant relation to the old English folktale of the Little Red Hen, but its moral is different.

5. The voivodeship (*województwo* in Polish) is a Polish administrative unit. Since 1999, Poland has sixteen voivodeships.

6. This is an oblique reference to a line from the historiosophical drama *Lilla Weneda* by the Polish Romantic poet Juliusz Słowacki (1809–1849): "No time to pity roses when a forest is on fire."

7. The Latin term *ad hoc* (literally "for this") denotes an explanation or solution to a concrete problem that is not subject to generalization or adaptation to other purposes. Although its connotation does not always have to be pejorative or restrictive, it most often labels solutions that are tentative and carry an element of improvisation and the absence of a plan. "*Ad usum Delphini*" ("for Dauphin's use") were the words stamped on the title pages of the text collections (including Homer, Ovid, Racin, and others) that were for education of the Grand Dauphin, which was the title of the French prince, the son of King Louis XIV (1638–1715). The classical texts were, however, "cleaned" of things that may have "embarrassed" the little prince or were otherwise "inappropriate." Today this term serves as a pejorative label for a text that has been censored before public distribution.

Supplement 2: Additional Essays

The Ethics of Technology and the Technology of Ethics

1. Jean Rostand (1894–1977) was a French biologist and philosopher. His main occupation was science popularization and history (as well as its future). Lem is probably referring to the book *Is It Possible to Modify the Human?* (J. Rostand, *Peut-on modifier l'Homme?* Paris, France: Gallimard, 1956). By the way, Jean's father Edmond was the dramatist who wrote the famous play *Cyrano de Bergerac* in 1897.

2. While spiders are related to insects, they are not them. Spiders and insects are two different classes of the phylum Arthropoda (along with scorpions, crabs, and shrimp). In contrast with insects, whose body consists of three parts—head, thorax, and abdomen—with six legs, and which, most often, can fly, a spider's body has only two segments—cephalothorax and abdomen—with eight legs, and they cannot fly. Nevertheless, this taxonomical detail does not affect the point of Lem's argument.

3. Recent professional literature confirms that LSD perturbs the recognition of danger and the psychological processing of fear, and suggests that LSD can find application in psychotherapy, where it improves sociability and increases empathy (see, e.g., P. C. Dolder, Y. Schmid, F. Müller, S. Borgwardt, and M. E. Liechti, "LSD acutely impairs fear recognition and enhances emotional empathy and sociality." Neuropsychopharmacology 41 (2016): 2638–2646, available on the website https://www.nature.com/articles/npp201682, accessed May 29, 2020).

4. Pangloss is a character from the satirical novel *Candide* by the French enlightenment philosopher François-Marie Arouet, known as Voltaire (1694-1778) (F. M. A. Voltaire, *Candide, ou L'Optimisme* [Paris, France: Sirène, 1759]). Professor Pangloss, a mentor of the protagonist Candide, supported the thesis that in this world, which is the best of all possible worlds, everything will end up well. This is a satire version of the philosophical optimism of the German rationalist Gottfried W. Leibniz (1646–1716).

5. The Mousterian culture flourished in Europe roughly between 160,000 and 40,000 years ago, and its carriers were apparently the Neanderthals. The name derives from the hamlet Peyzac-le-Moustier in the département Dordogne in France. The Aurignacian culture in Europe is younger, dating from 48,000 to 25,000 years ago, and its carrier was already *Homo sapiens sapiens*. It received its name from the village Aurignac in the département Haute-Garonne in the southwest of France.

6. Vladimir I. Vernadsky (1863–1945) was a Russian Ukrainian mineralogist and geochemist who was the first to divide the planet Earth into three spheres that influence each other: the geosphere, in which the geological processes take place, the biosphere, which is the domain of the life process, and the noosphere, where the dominant role is played by the human, his intellect, and the related activities. His seminal work was the book *The Biosphere* (V. I. Vernadsky, *Biosfera*. Leningrad, USSR: Naučn. kimi.-techn. izdateľstvo, 1926). Thus, Vernadsky was a predecessor of James Lovelock and Lynn Margulis, who later coind the concept that the Earth as a planet represents a unified homeostatic living system that they named Gea or, in an alternative transliteration, Gaia (the ancient Greeks used this name for the personification of the Earth).

7. O. S. Kulagina, "K voprosu o modelirovanii evoľucionnogo processa," in *Voprosy kibernetiki, vypusk 16*, ed. O. S. Kulagina and A. A. Lyapunov (Moscow: Nauka, 1966), 147–169. (The English text is my rendition of Lem's Polish translation of the Russian original.)

8. Orthoevolution or progressive evolution comes from the term orthogenesis (from Ancient Greek ὀρθός, "straight," and γένεσις, "origin"), which was first used by the biologist Wilhelm Haacke in 1893. Ethymologically, orthoevolution can be understood as "an evolution in a straight line," but the term's scientific content has been very elastic throughout its history, as it meant different things to different people. Often, mainly close to its origin, it was considered an alternative to Darwinism because it gave evolution a kind of preset goal and even the tracks for achieving it, independent of genetic variability and natural selection. In the middle of the twentieth century, Darwinists like Ernst Mayr declared it outright nonscientific due to its (possible) ties to teleology, evolutionary "progress," or even the so-called vital force, an ancient undefined, mystical parameter that was supposed to be the difference between animate and inanimate objects. However, orthoevolution has never left the field of evolutionary biology entirely, even though some still view it as "heresy." Apparently, it is a "popular" heresy since it is present, in more or less veiled forms, in the views of prominent representatives of the Darwinian "modern synthesis," such as E. O. Wilson or R. Dawkins. In 1996, the philosopher of biology Michael Ruse defined orthogenesis as "the view that evolution has a kind of momentum of its own that carries organisms along certain tracks" (Michael Ruse, *Monad to Man: The Concept of Progress in Evolutionary Biology* [Cambridge, MA: Harvard University Press, 1996]). Of course, orthoevolution's biggest problem is the identity of this "momentum," which gives the "direction" to the evolutionary "tracks." Lem uses the term without apprehension and furthermore, seems to suggest that this momentum might be the inherent tendency of complex homeostatic systems to increase their complexity and diversity, whose roots transcend biology because they are found in nonequilibrium thermodynamics and the complex-system science.

9. J. Willard Gibbs (1839–1904) was a scientist whom many (including A. Einstein) considered the best in the history of the United States. As a theorist, he made essential contributions to the development of physics, chemistry, and mathematics. Together with James C. Maxwell and Ludwig Boltzmann, he created a new field of thermodynamics—statistical thermodynamics.

10. Rudolf Carnap (1891–1970) was a German American philosopher, one of the most prominent members of the Vienna circle and a proponent of logical positivism, whose aim was to make the "pure" philosophy a true "scientific discipline." The discussion of the scientific and ethical value of the statement about killing comes from his perhaps most famous book *Philosophy and Logical Syntax* (London: Kegan Paul, Trench, Trubner and Co., 1935).

11. Lem is referring to perhaps the best known quote of the French mathematician, physicist, and philosopher Blaise Pascal (1623–1662) from his *Thoughts* (B. Pascal, *Pensées* [Paris, France: Guillaume Desprez, 1670]): "Man is but a reed, the most feeble thing in nature; but he is a thinking reed. The entire universe need not arm itself to crush him. A vapor, a drop of water suffices to kill him. But, if the universe were to crush him, man would still be more noble than that which killed him, because he knows that he dies and the advantage that the universe has over him; the universe knows nothing of this."

Biology and Values

1. Hypostasis (from the Greek ὑπόστασις, literally "a state that is underneath") in Neoplatonic philosophy denotes the essence or the fundamental reality hidden underneath the surface, but in descriptive linguistics it refers to using a synsemantic word (a word that has meaning only in the context provided by other, surrounding words) as if it were autosemantic (a word that has meaning regardless of the context). Conceptually, linguistic hypostasis can be understood as a process that is the opposite of generalizing an originally specific term, such as a proper name or a brand name, to denote all objects of a given kind.

2. "End of the world" (Lat.).

3. "Albedo" and "insolation" are physical terms denoting the reflection of radiation by a surface and the amount of radiative energy emitted into space.

4. "Superfluous entitites" (Lat.). It is an allusion to Ockham's razor.

5. F. Rosenblatt, *Principles of Neurodynamics: Perceptrons and the Theory of Brain Mechanisms* (Washington, DC: Spartan, 1962).

6. "Speciation" refers to the emergence of a species.

7. The cardinality of a set is a technical term that generalizes the notion of the number of elements in the set. In the case of a finite set, cardinality is simply equal to the number of its elements.

8. Stafford Beer (1926–2002) was a British theorist and practitioner in operational research and management cybernetization. The cited book, *Cybernetics and Management* (London: English Universities Press, 1959), is his most acclaimed work.

9. Finitism is a philosophy of mathematics that allows the existence of only finite mathematical objects.

10. DNA is an abbreviation for deoxyribonucleic acid, a molecule that carries the genetic information. The molecule is a linear polymer that is very tightly packed in the chromosomes in the cell's nucleus; when extended, the total length of DNA in each human cell is over one meter.

11. The word "entelechy" (from the Old Greek ἐντελέχεια) is probably an invention of Aristotle (384–322 BCE). He composed it from the words ἐντελής (completed, whole, perfect) and ἔχω (to have). In his metaphysics, it denoted the complete realization of the final form of a potential concept or function. In modern general philosophy it can denote

the motivation or the need for autonomy and, at the same time, the inner strength for its actualization.

12. "Logical depth" is a technical term denoting the measure of a system's complexity, which was proposed by the US computer scientist and mathematician Charles Bennett (b. 1942). It is based on the so-called computational approach to complexity: logical depth equals the time that the shortest possible algorithm needs to create the complete model of the given system.

13. This is a biochemical misunderstanding. Cells need oxygen not as a source of electrons but, on the contrary, as their final acceptor. The electrons, from which energy is extracted in the cellular respiratory chain (through the coupling with oxidative phosphorylation), originate from the molecules of foodstuff and end up on the molecule of oxygen (O_2), which is thereby reduced (in the chemical sense) and, in the form of carbon dioxide (CO_2) and water (H_2O), is returned to the environment. Indeed, each atom of oxygen in the molecules that we exhale is two electrons richer than in the molecules of oxygen that we inhale. Hence, the "electron packets" would not replace the breathing of oxygen but rather the eating of food. Lem probably meant that we could invent an "electron acceptor" other than oxygen.

14. Tracheae comprise a system of narrowing tubes that branch throughout the whole body of insects. The tubes, strengthened by chitin, directly but passively distribute the air (with oxygen) to all the insect's tissues.

15. The allele (from the Greek ἄλλος, "another") is one of the alternative forms of a gene. The gene usually has two alleles—one from the mother and the other from the father. However, the "strength" of the two alleles is not necessarily the same. The "stronger" one is called dominant. When the dominant allele appears in a given gene, the organism will always manifest the feature that this allele encodes, even when it is paired with a different allele that is "weaker" (called recessive) from the other parent. Hence, the feature encoded by a recessive allele will not manifest in the given organism (in its phenotype), but the organism continues to be its carrier and can transmit it to its progeny. The trait encoded by a recessive allele can thus be transmitted for many generations without notice and it will only manifest in the phenotype when it pairs with another recessive allele (from the other parent) in the given "locus" of the gene.

16. The American psychologist John. B. Watson (1878–1958) founded the behavioral school of psychology in his lecture at the Columbia University in 1913 (see J. B. Watson, "Psychology as the Behaviorist Views It," *Psychological Review* 20 (1913): 158–177). Behaviorism pushes the so-called depth psychology with its non-empirical interpretations into the background. It treats behavior as a set of observable physical phenomena and presumes that each of its elements has a physical cause that can be objectively traced. While it cannot be said that behaviorism is dead today, it was pushed out of the forefront in the second half of the twentieth century by a new approach, cognitive psychology, which can be considered a modern compromise between the "physicalistic" behaviorism and the previous "mentalistic" psychology—with the significant contribution of biomolecular research techniques.

17. In Latin, "nothing can arise out of nothing." This statement is ascribed to the most important philosopher of the Eleatic school, Parmenides (sixth century BCE), who applied it to the existence itself: according to him, being is eternal, and what is, must have always been, because nothing can arise out of nothing.

18. Ecclesiastes 9:4.

19. "Vitreous" freezing produces amorphous, vitreous, or noncrystalline ice, which does not have a lower density (a larger volume) than liquid water and therefore does not damage the cell or tissue that is being frozen. It is achieved by a very rapid freezing of microscopic and thin-sectioned specimens. Because the bulkier, macroscopic tissues or organs inevitably freeze more slowly (due to the physics of heat propagation), regular ice crystals that cause irreversible cell damage form in their interior .

20. Julian S. Huxley (1887–1975) was a British evolutionary biologist and science popularizer, who believed in the objectivity of evolutionary progress, even though he admitted that evolution has no goal. His most known work is *Evolution: The Modern Synthesis* (London: George Allen and Unwin, 1942). Huxley hailed from a prominent intellectual family: his grandfather Thomas H. Huxley was a friend and a supporter of Charles Darwin; his younger brother Aldous Huxley gained fame for his dystopic novel *Brave New World*; and their still younger stepbrother Andrew Huxley received the Nobel Prize (together with Alan Hodgkin) for the discovery of the action potential through which electric impulses propagate in the nerve cell.

21. Individuation is the process of becoming oneself. This concept plays a prominent role in the thought systems of many past (Bergson, Nietzsche, and Jung) and contemporary (Deleuze, Stiegler, Simondon) philosophers.